Grouping Multidimensional Data

Jacob Kogan · Charles Nicholas
Marc Teboulle (Eds.)

Grouping Multidimensional Data

Recent Advances in Clustering

With 53 Figures

 Springer

Editors

Jacob Kogan

Department of Mathematics and Statistics
and Department of Computer Science
and Electrical Engineering
University of Maryland Baltimore County
1000 Hilltop Circle
Baltimore, Maryland 21250, USA
kogan@umbc.edu

Marc Teboulle

School of Mathematical Sciences
Tel-Aviv University
Ramat Aviv, Tel-Aviv 69978, Israel
teboulle@post.tau.ac.il

Charles Nicholas

Department of Computer Science
and Electrical Engineering
University of Maryland Baltimore County
1000 Hilltop Circle
Baltimore, Maryland 21250, USA
nicholas@umbc.edu

ACM Classification (1998): H.3.1, H.3.3
Library of Congress Control Number: 2005933258

ISBN-10 3-540-28348-X Springer Berlin Heidelberg New York
ISBN-13 978-3-540-28348-5 Springer Berlin Heidelberg New York

This work is subject to copyright. All rights are reserved, whether the whole or part of the material is concerned, specifically the rights of translation, reprinting, reuse of illustrations, recitation, broadcasting, reproduction on microfilm or in any other way, and storage in data banks. Duplication of this publication or parts thereof is permitted only under the provisions of the German Copyright Law of September 9, 1965, in its current version, and permission for use must always be obtained from Springer. Violations are liable for prosecution under the German Copyright Law.

Springer is a part of Springer Science+Business Media
springeronline.com
© Springer-Verlag Berlin Heidelberg 2006
Printed in The Netherlands

The use of general descriptive names, registered names, trademarks, etc. in this publication does not imply, even in the absence of a specific statement, that such names are exempt from the relevant protective laws and regulations and therefore free for general use.

Typesetting by the authors and SPI Publisher Services using Springer LATEX macro package

Cover design: KünkelLopka, Heidelberg

Printed on acid-free paper   SPIN: 11375456   45/3100/ SPI Publisher Services   5 4 3 2 1 0

Foreword

Clustering is one of the most fundamental and essential data analysis tasks with broad applications. It can be used as an independent data mining task to disclose intrinsic characteristics of data, or as a preprocessing step with the clustering results used further in other data mining tasks, such as classification, prediction, correlation analysis, and anomaly detection. It is no wonder that clustering has been studied extensively in various research fields, including data mining, machine learning, pattern recognition, and scientific, engineering, social, economic, and biomedical data analysis. Although there have been numerous studies on clustering methods and their applications, due to the wide spectrum that the theme covers and the diversity of the methodology research publications on this theme have been scattered in various conference proceedings or journals in multiple research fields. There is a need for a good collection of books dedicated to this theme, especially considering the surge of research activities on cluster analysis in the last several years.

This book fills such a gap and meets the demand of many researchers and practitioners who would like to have a solid grasp of the state of the art on cluster analysis methods and their applications. The book consists of a collection of chapters, contributed by a group of authoritative researchers in the field. It covers a broad spectrum of the field, from comprehensive surveys to in-depth treatments of a few important topics. The book is organized in a systematic manner, treating different themes in a balanced way. It is worth reading and further when taken as a good reference book on your shelf.

The chapter "A Survey of Clustering Data Mining Techniques" by Pavel Berkhin provides an overview of the state-of-the-art clustering techniques. It presents a comprehensive classification of clustering methods, covering hierarchical methods, partitioning relocation methods, density-based partitioning methods, grid-based methods, methods based on co-occurrence of categorical data, and other clustering techniques, such as constraint-based and graph-partitioning methods. Moreover, it introduces scalable clustering algorithms and clustering algorithms for high-dimensional data. Such a coverage provides a well-organized picture of the whole research field.

In the chapter "Similarity-Based Text Clustering: A Comparative Study," Joydeep Ghosh and Alexander Strehl perform the first comparative study among popular similarity measures (Euclidean, cosine, Pearson correlation, extended Jaccard) in conjunction with several clustering techniques (random, self-organizing feature map, hypergraph partitioning, generalized k-means, weighted graph partitioning) on a variety of high-dimensional sparse vector data sets representing text documents as bags of words. The comparative performance results are interesting and instructive.

In the chapter "Criterion Functions for Clustering on High-Dimensional Data", Ying Zhao and George Karypis provide empirical and theoretical comparisons of the performance of a number of widely used criterion functions in the context of partitional clustering algorithms for high-dimensional datasets. This study presents empirical and theoretical guidance on the selection of criterion functions for clustering high-dimensional data, such as text documents.

Other chapters also provide interesting introduction and in-depth treatments of various topics of clustering, including a star-clustering algorithm by Javed Aslam, Ekaterina Pelekhov, and Daniela Rus, a study on clustering large datasets with principal direction divisive partitioning by David Littau and Daniel Boley, a method for clustering with entropy-like k-means algorithms by Marc Teboulle, Pavel Berkhin, Inderjit Dhillon, Yuqiang Guan, and Jacob Kogan, two new sampling methods for building initial partitions for effective clustering by Zeev Volkovich, Jacob Kogan, and Charles Nicholas, and "TMG: A MATLAB Toolbox for Generating Term-Document Matrices from Text Collections" by Dimitrios Zeimpekis and Efstratios Gallopoulos. These chapters present in-depth treatment of several popularly studied methods and widely used tools for effective and efficient cluster analysis.

Finally, the book provides a comprehensive bibliography, which is a marvelous and up-to-date list of research papers on cluster analysis. It serves as a valuable resource for researchers.

I enjoyed reading the book. I hope you will also find it a valuable source for learning the concepts and techniques of cluster analysis and a handy reference for in-depth and productive research on these topics.

University of Illinois at   Jiawei Han
Urbana-Champaign
June 29, 2005

Preface

Clustering is a fundamental problem that has numerous applications in many disciplines. Clustering techniques are used to discover natural groups in datasets and to identify abstract structures that might reside there, without having any background knowledge of the characteristics of the data. They have been used in various areas including bioinformatics, computer vision, data mining, gene expression analysis, text mining, VLSI design, and Web page clustering to name just a few. Numerous recent contributions to this research area are scattered in a variety of publications in multiple research fields.

This volume collects contributions of computers scientists, data miners, applied mathematicians, and statisticians from academia and industry. It covers a number of important topics and provides about 500 references relevant to current clustering research (we plan to make this reference list available on the Web). We hope the volume will be useful for anyone willing to learn about or contribute to clustering research.

The editors would like to express gratitude to the authors for making their research available for the volume. Without these individuals' help and cooperation this book would not be possible. Thanks also go to Ralf Gerstner of Springer for his patience and assistance, and for the timely production of this book. We would like to acknowledge the support of the United States–Israel Binational Science Foundation through the grant BSF No. 2002-010, and the support of the Fulbright Program.

Karmiel, Israel and Baltimore, USA,   Jacob Kogan
Baltimore, USA,   Charles Nicholas
Tel Aviv, Israel,   Marc Teboulle
July 2005

Contents

List of Contributors

J. A. Aslam
College of Computer and
Information Science
Northeastern University
Boston, MA 02115, USA
jaa@ccs.neu.edu

P. Berkhin
Yahoo!
701 First Avenue
Sunnyvale, CA 94089, USA
pberkhin@yahoo-inc.com

D. Boley
University of Minnesota
Minneapolis, MN 55455, USA
boley@cs.umn.edu

I. Dhillon
Department of Computer Science
University of Texas
Austin, TX 78712-1188, USA
inderjit@cs.utexas.edu

E. Gallopoulos
Department of Computer
Engineering and Informatics
University of Patras
26500 Patras
Greece
stratis@hpclab.ceid.upatras.gr

J. Ghosh
Department of ECE
University of Texas at Austin
1 University Station C0803
Austin, TX 78712-0240, USA
ghosh@ece.utexas.edu

Y. Guan
Department of Computer Science
University of Texas
Austin, TX 78712-1188, USA
yguan@cs.utexas.edu

G. Karypis
Department of Computer Science
and Engineering and Digital
Technology Center and
Army HPC Research Center
University of Minnesota
Minneapolis, MN 55455, USA
karypis@cs.umn.edu

J. Kogan
Department of Mathematics and
Statistics and
Department of Computer Science
and Electrical Engineering
University of Maryland
Baltimore County
Baltimore, MD 21250, USA
kogan@umbc.edu

D. Littau
University of Minnesota
Minneapolis, MN 55455, USA
littau@cs.umn.edu

C. Nicholas
Department of Computer Science
and Electrical Engineering
University of Maryland
Baltimore County
Baltimore, MD 21250, USA
nicholas@csee.umbc.edu

E. Pelekhov
Department of Computer Science
Dartmouth College
Hanover, NH 03755, USA
ekaterina.pelekhov@alum.
dartmouth.org

D. Rus
Computer Science and Artificial
Intelligence Laboratory
Massachusetts Institute of
Technology
Cambridge, MA 02139, USA
rus@csail.mit.edu

A. Strehl
Leubelfingstrasse 110
90431 Nurnberg
Germany
alexander@strehl.com

M. Teboulle
School of Mathematical Sciences
Tel Aviv University
Tel Aviv, Israel
teboulle@post.tau.ac.il

Z. Volkovich
Software Engineering Department
ORT Braude Academic College
Karmiel 21982, Israel
zeev@actcom.co.il

D. Zeimpekis
Department of Computer
Engineering and Informatics
University of Patras
26500 Patras
Greece
dsz@hpclab.ceid.upatras.gr

Y. Zhao
Department of Computer Science
and Engineering
University of Minnesota
Minneapolis, MN 55455, USA
yzhao@cs.umn.edu

The Star Clustering Algorithm for Information Organization

J.A. Aslam, E. Pelekhov, and D. Rus

Summary. We present the star clustering algorithm for static and dynamic information organization. The offline star algorithm can be used for clustering static information systems, and the online star algorithm can be used for clustering dynamic information systems. These algorithms organize a data collection into a number of clusters that are naturally induced by the collection via a computationally efficient cover by dense subgraphs. We further show a lower bound on the accuracy of the clusters produced by these algorithms as well as demonstrate that these algorithms are computationally efficient. Finally, we discuss a number of applications of the star clustering algorithm and provide results from a number of experiments with the Text Retrieval Conference data.

1 Introduction

We consider the problem of automatic information organization and present the star clustering algorithm for static and dynamic information organization. Offline information organization algorithms are useful for organizing static collections of data, for example, large-scale legacy collections. Online information organization algorithms are useful for keeping dynamic corpora, such as news feeds, organized. Information retrieval (IR) systems such as Inquery [427], Smart [378], and Google provide automation by computing ranked lists of documents sorted by relevance; however, it is often ineffective for users to scan through lists of hundreds of document titles in search of an information need. Clustering algorithms are often used as a preprocessing step to organize data for browsing or as a postprocessing step to help alleviate the "information overload" that many modern IR systems engender.

There has been extensive research on clustering and its applications to many domains [17, 231]. For a good overview see [242]. For a good overview of using clustering in IR see [455]. The use of clustering in IR was mostly driven by *the cluster hypothesis* [429], which states that "closely associated documents tend to be related to the same requests." Jardine and van Rijsbergen [246] show some evidence that search results could be improved by clustering. Hearst and Pedersen [225] re-examine the cluster hypothesis by focusing on the Scatter/Gather system [121] and conclude that it holds for browsing tasks.

Systems like Scatter/Gather [121] provide a mechanism for user-driven organization of data in a fixed number of clusters but the users need to be in the loop and the computed clusters do not have accuracy guarantees. Scatter/Gather uses fractionation to compute nearest-neighbor clusters. Charika et al. [104] consider a dynamic clustering algorithm to partition a collection of text documents into a *fixed* number of clusters. Since in dynamic information systems the number of topics is not known a priori, a fixed number of clusters cannot generate a natural partition of the information.

In this chapter, we provide an overview of our work on clustering algorithms and their applications [26–33]. We propose an offline algorithm for clustering static information and an online version of this algorithm for clustering dynamic information. These two algorithms compute clusters induced by the natural topic structure of the information space. Thus, this work is different from [104, 121] in that we do not impose the constraint to use a fixed number of clusters. As a result, we can guarantee a lower bound on the topic similarity between the documents in each cluster. The model for topic similarity is the standard vector space model used in the IR community [377], which is explained in more detail in Sect. 2 of this chapter.

While the clustering document represented in the vector space model is our primary motivating example, our algorithms can be applied to clustering any set of objects for which a similarity measure is defined, and the performance results stated largely apply whenever the objects themselves are represented in a feature space in which similarity is defined by the cosine metric.

To compute accurate clusters, we formalize clustering as covering graphs by cliques [256] (where the cover is a vertex cover). Covering by cliques is NP complete and thus intractable for large document collections. Unfortunately, it has also been shown that the problem cannot be approximated even in polynomial time [322, 465]. We instead use a cover by *dense subgraphs* that are *star shaped* and that can be computed *offline* for static data and *online* for dynamic data. We show that the offline and the online algorithms produce correct clusters efficiently. Asymptotically, the running time of both algorithms is roughly linear in the size of the similarity graph that defines the information space (explained in detail in Sect. 2). We also show lower bounds on the topic similarity within the computed clusters (a measure of the accuracy of our clustering algorithm) as well as provide experimental data.

We further compare the performance of the star algorithm to two widely used algorithms for clustering in IR and other settings: the single link method[1] [118] and the average link algorithm[2] [434]. Neither algorithm provides guarantees for the topic similarity within a cluster. The single link algorithm can be used in offline and online modes, and it is faster than the average link algorithm, but it produces poorer clusters than the average link algorithm. The average link algorithm can only be used offline to process static data. The star clustering algorithm, on the other hand, computes topic clusters that are naturally induced by the collection, provides guarantees on cluster quality, computes more accurate clusters than either the single link or the average link methods, is efficient, admits an efficient and simple online version, and can perform hierarchical data organization. We describe experiments in this chapter with the TREC[3] collection demonstrating these abilities.

Finally, we discuss the use of the star clustering algorithm in a number of different application areas including (1) automatic information organization systems, (2) scalable information organization for large corpora, (3) text filtering, and (4) persistent queries.

2 Motivation for the Star Clustering Algorithm

In this section we describe our clustering model and provide motivation for the star clustering algorithm. We begin by describing the vector space model for document representation and consider an idealized clustering algorithm based on clique covers. Given that clique cover algorithms are computationally infeasible, we redundant propose an algorithm based on star covers. Finally, we argue that star covers retain many of the desired properties of clique covers in expectation, and we demonstrate in subsequent sections that clusterings based on star covers can be computed very efficiently both online and offline.

2.1 Clique Covers in the Vector Space Model

We formulate our problem by representing a document collection by its *similarity graph*. A similarity graph is an undirected, weighted graph $G = (V, E, w)$, where the vertices in the graph correspond to documents and each weighted edge in the graph corresponds to a measure of similarity between two documents. We measure the similarity between two documents by using a standard metric from the IR community – the cosine metric in the vector space model of the Smart IR system [377, 378].

---

[1] In the single link clustering algorithm a document is part of a cluster if it is "related" to at least *one* document in the cluster

[2] In the average link clustering algorithm a document is part of a cluster if it is "related" to an average number of documents in the cluster

[3] TREC is the Annual Text Retrieval Conference. Each participant is given of the order of 5 GB of data and a standard set of queries to test the systems. The results and the system descriptions are presented as papers at the TREC

The vector space model for textual information aggregates statistics on the occurrence of words in documents. The premise of the vector space model is that two documents are similar if they use similar words. A vector space can be created for a collection (or corpus) of documents by associating each important word in the corpus with one dimension in the space. The result is a high-dimensional vector space. Documents are mapped to vectors in this space according to their word frequencies. Similar documents map to nearby vectors. In the vector space model, document similarity is measured by the angle between the corresponding document vectors. The standard in the IR community is to map the angles to the interval $[0, 1]$ by taking the cosine of the vector angles.

G is a complete graph with edges of varying weight. An organization of the graph that produces reliable clusters of similarity σ (i.e., clusters where documents have pairwise similarities of at least σ) can be obtained by (1) thresholding the graph at σ and (2) performing a *minimum clique cover* with maximal cliques on the resulting graph G_σ. The *thresholded graph* G_σ is an undirected graph obtained from G by eliminating all the edges whose weights are lower than σ. The minimum clique cover has two features. First, by using cliques to cover the similarity graph, we are guaranteed that all the documents in a cluster have the desired degree of similarity. Second, minimal clique covers with maximal cliques allow vertices to belong to *several* clusters. In many information retrieval applications, this is a desirable feature as documents can have multiple subthemes.

Unfortunately, this approach is computationally intractable. For real corpora, similarity graphs can be very large. The clique cover problem is NP-complete, and it does not admit polynomial-time approximation algorithms [322, 465]. While we cannot perform a clique cover or even approximate such a cover, we can instead cover our graph by *dense subgraphs*. What we lose in intracluster similarity guarantees, we gain in computational efficiency.

2.2 Star Covers

We approximate a clique cover by covering the associated thresholded similarity graph with *star-shaped subgraphs*. A star-shaped subgraph on $m+1$ vertices consists of a single *star center* and m *satellite vertices*, where there exist edges between the star center and each of the satellite vertices (see Fig. 1). While finding cliques in the thresholded similarity graph G_σ guarantees a pairwise similarity between documents of at least σ, it would appear at first glance that finding star-shaped subgraphs in G_σ would provide similarity guarantees between the star center and each of the satellite vertices, but no such similarity guarantees *between satellite vertices*. However, by investigating the geometry of our problem in the vector space model, we can derive a *lower bound* on the similarity between satellite vertices as well as provide a formula for the *expected* similarity between satellite vertices. The latter formula predicts that the pairwise similarity between satellite vertices in a star-shaped subgraph is

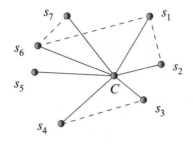

Fig. 1. An example of a star-shaped subgraph with a center vertex C and satellite vertices s_1–s_7. The edges are denoted by solid and dashed lines. Note that there is an edge between each satellite and a center, and that edges may also exist between satellite vertices

high, and together with empirical evidence supporting this formula, we conclude that covering G_σ with star-shaped subgraphs is an accurate method for clustering a set of documents.

Consider three documents C, s_1, and s_2 that are vertices in a star-shaped subgraph of G_σ, where s_1 and s_2 are satellite vertices and C is the star center. By the definition of a star-shaped subgraph of G_σ, we must have that the similarity between C and s_1 is at least σ and that the similarity between C and s_2 is also at least σ. In the vector space model, these similarities are obtained by taking the cosine of the angle between the vectors associated with each document. Let α_1 be the angle between C and s_1, and let α_2 be the angle between C and s_2. We then have that $\cos \alpha_1 \geq \sigma$ and $\cos \alpha_2 \geq \sigma$. Note that the angle between s_1 and s_2 can be at most $\alpha_1 + \alpha_2$; we therefore have the following lower bound on the similarity between satellite vertices in a star-shaped subgraph of G_σ.

Theorem 1. *Let G_σ be a similarity graph and let s_1 and s_2 be two satellites in the same star in G_σ. Then the similarity between s_1 and s_2 must be at least*

$$\cos(\alpha_1 + \alpha_2) = \cos \alpha_1 \cos \alpha_2 - \sin \alpha_1 \sin \alpha_2.$$

The use of Theorem 1 to bound the similarity between satellite vertices can yield somewhat disappointing results. For example, if $\sigma = 0.7$, $\cos \alpha_1 = 0.75$, and $\cos \alpha_2 = 0.85$, we can conclude that the similarity between the two satellite vertices must be at least[4]:

$$0.75 \times 0.85 - \sqrt{1 - (0.75)^2}\sqrt{1 - (0.85)^2} \approx 0.29.$$

---

[4]Note that $\sin \theta = \sqrt{1 - \cos^2 \theta}$

Note that while this may not seem very encouraging, the analysis is based on absolute worst-case assumptions, and in practice, the similarities between satellite vertices are much higher. We can instead reason about the *expected* similarity between two satellite vertices by considering the geometric constraints imposed by the vector space model as follows.

Theorem 2. *Let C be a star center, and let S_1 and S_2 be the satellite vertices of C. Then the similarity between S_1 and S_2 is given by*

$$\cos \alpha_1 \cos \alpha_2 + \cos \theta \sin \alpha_1 \sin \alpha_2,$$

where θ is the dihedral angle[5] between the planes formed by S_1C and S_2C.

This theorem is a fairly direct consequence of the geometry of C, S_1, and S_2 in the vector space; details may be found in [31].

How might we eliminate the dependence on $\cos \theta$ in this formula? Consider three vertices from a cluster of similarity σ. Randomly chosen, the pairwise similarities among these vertices should be $\cos \omega$ for some ω satisfying $\cos \omega \geq \sigma$. We then have

$$\cos \omega = \cos \omega \cos \omega + \cos \theta \sin \omega \sin \omega$$

from which it follows that

$$\cos \theta = \frac{\cos \omega - \cos^2 \omega}{\sin^2 \omega} = \frac{\cos \omega (1 - \cos \omega)}{1 - \cos^2 \omega} = \frac{\cos \omega}{1 + \cos \omega}.$$

Substituting for $\cos \theta$ and noting that $\cos \omega \geq \sigma$, we obtain

$$\cos \gamma \geq \cos \alpha_1 \cos \alpha_2 + \frac{\sigma}{1 + \sigma} \sin \alpha_1 \sin \alpha_2. \tag{1}$$

Equation (1) provides an accurate estimate of the similarity between two satellite vertices, as we demonstrate empirically.

Note that for the example given in Sect. 2.2, (1) would predict a similarity between satellite vertices of approximately 0.78. We have tested this formula against real data, and the results of the test with the TREC FBIS data set[6] are shown in Fig. 2. In this plot, the x-axis and y-axis are similarities between cluster centers and satellite vertices, and the z-axis is the root mean squared prediction error (RMS) of the formula in Theorem 2 for the similarity between satellite vertices. We observe the maximum root mean squared error is quite small (approximately 0.16 in the worst case), and for reasonably high similarities, the error is negligible. From our tests with real data, we have concluded that (1) is quite accurate. We may further conclude that star-shaped subgraphs are reasonably "dense" in the sense that they imply relatively high pairwise similarities between all documents in the star.

---

[5] The dihedral angle is the angle between two planes on a third plane normal to the intersection of the two planes

[6] Foreign Broadcast Information Service (FBIS) is a large collection of text documents used in TREC

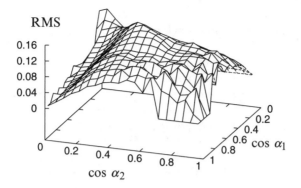

Fig. 2. The RMS prediction error of our expected satellite similarity formula over the TREC FBIS collection containing 21,694 documents

3 The Offline Star Clustering Algorithm

Motivated by the discussion of Sect. 2, we now present the *star algorithm*, which can be used to organize documents in an information system. The star algorithm is based on a greedy cover of the thresholded similarity graph by star-shaped subgraphs; the algorithm itself is summarized in Fig. 3 below.

Theorem 3. *The running time of the offline star algorithm on a similarity graph G_σ is $\Theta(V + E_\sigma)$.*

For any threshold σ:

1. Let $G_\sigma = (V, E_\sigma)$ where $E_\sigma = \{e \in E : w(e) \geq \sigma\}$.
2. Let each vertex in G_σ initially be *unmarked*.
3. Calculate the degree of each vertex $v \in V$.
4. Let the highest degree unmarked vertex be a star center, and construct a cluster from the star center and its associated satellite vertices. Mark each node in the newly constructed star.
5. Repeat Step 4 until all nodes are marked.
6. Represent each cluster by the document corresponding to its associated star center.

Fig. 3. The star algorithm

Proof. The following implementation of this algorithm has a running time *linear* in the size of the graph. Each vertex v has a data structure associated with it that contains $v.degree$, the degree of the vertex, $v.adj$, the list of adjacent vertices, $v.marked$, which is a bit denoting whether the vertex belongs to a star or not, and $v.center$, which is a bit denoting whether the vertex is a star center. (Computing $v.degree$ for each vertex can be easily performed in $\Theta(V + E_\sigma)$ time.) The implementation starts by sorting the vertices in V by degree ($\Theta(V)$ time since degrees are integers in the range $\{0, |V|\}$). The program then scans the sorted vertices from the highest degree to the lowest as a greedy search for star centers. Only vertices that do not belong to a star already (that is, they are unmarked) can become star centers. Upon selecting a new star center v, its $v.center$ and $v.marked$ bits are set and for all $w \in v.adj$, $w.marked$ is set. Only one scan of V is needed to determine all the star centers. Upon termination, the star centers and only the star centers have the *center* field set. We call the set of star centers the *star cover* of the graph. Each star is fully determined by the star center, as the satellites are contained in the adjacency list of the center vertex. □

This algorithm has two features of interest. The first feature is that the star cover is not unique. A similarity graph may have several different star covers because when there are several vertices of the same highest degree, the algorithm arbitrarily chooses one of them as a star center (whichever shows up first in the sorted list of vertices). The second feature of this algorithm is that it provides a simple encoding of a star cover by assigning the types "center" and "satellite" (which is the same as "not center" in our implementation) to vertices. We define a *correct star cover* as a star cover that assigns the types "center" and "satellite" in such a way that (1) a star center is not adjacent to any other star center and (2) every satellite vertex is adjacent to at least one center vertex of equal or higher degree.

Figure 4 shows two examples of star covers. The left graph consists of a clique subgraph (first subgraph) and a set of nodes connected to only to the nodes in the clique subgraph (second subgraph). The star cover of the left graph includes one vertex from the 4-clique subgraph (which covers the entire clique and the one nonclique vertex it is connected to), and single-node stars for each of the noncovered vertices in the second set. The addition of a node connected to all the nodes in the second set changes the clique cover dramatically. In this case, the new node becomes a star center. It thus covers all the nodes in the second set. Note that since star centers cannot be adjacent, no vertex from the second set is a star center in this case. One node from the first set (the clique) remains the center of a star that covers that subgraph. This example illustrates the connection between a star cover and other important graph sets, such as set covers and induced dominating sets, which have been studied extensively in the literature [19, 183]. The star cover is related but not identical to a dominating set [183]. Every star cover is a dominating set, but there are dominating sets that are not star covers.

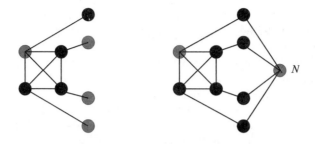

Fig. 4. An example of a star-shaped cover before and after the insertion of the node N in the graph. The dark circles denote satellite vertices. The shaded circles denote star centers

Star covers are useful approximations of clique covers because star graphs are dense subgraphs for which we can infer something about the missing edges as we have shown earlier.

Given this definition for the star cover, it immediately follows that:

Theorem 4. *The offline star algorithm produces a correct star cover.*

We use the two features of the offline algorithm mentioned earlier in the analysis of the online version of the star algorithm in Sect. 4. In Sect. 5, we show that the clusters produced by the star algorithm are quite accurate, exceeding the accuracy produced by widely used clustering algorithms in IR.

4 The Online Star Algorithm

The star clustering algorithm described in Sect. 3 can be used to accurately and efficiently cluster a static collection of documents. However, it is often the case in information systems that documents are added to, or deleted from, a dynamic collection. In this section, we describe an online version of the star clustering algorithm, which can be used to efficiently maintain a star clustering in the presence of document insertions and deletions.

We assume that documents are inserted or deleted from the collection one at a time. We begin by examining INSERT. The intuition behind the incremental computation of the star cover of a graph after a new vertex is inserted is depicted in Fig. 5. The top figure denotes a similarity graph and a correct star cover for this graph. Suppose a new vertex is inserted in the graph, as in the middle figure. The original star cover is no longer correct for the new graph. The bottom figure shows the correct star cover for the new graph. How does the addition of this new vertex affect the correctness of the star cover?

In general, the answer depends on the degree of the new vertex and its adjacency list. If the adjacency list of the new vertex does not contain any star centers, the new vertex can be added in the star cover as a star center. If the adjacency list of the new vertex contains any center vertex c whose degree is equal or higher, the new vertex becomes a satellite vertex of c. The difficult cases that destroy the correctness of the star cover are (1) when the new vertex is adjacent to a collection of star centers, each of whose degree is lower than that of the new vertex and (2) when the new vertex increases the degree of an adjacent satellite vertex beyond the degree of its associated star center. In these situations, the star structure already in place has to be modified; existing stars must be broken. The satellite vertices of these broken stars must be re-evaluated.

Similarly, deleting a vertex from a graph may destroy the correctness of a star cover. An initial change affects a star if (1) its center is removed or (2) the degree of the center has decreased because of a deleted satellite. The satellites in these stars may no longer be adjacent to a center of equal or higher degree, and their status must be reconsidered.

4.1 The Online Algorithm

Motivated by the intuition in the previous section, we now describe a simple online algorithm for incrementally computing star covers of dynamic graphs. The algorithm uses a data structure to efficiently maintain the star covers of an undirected graph $G = (V, E)$. For each vertex $v \in V$, we maintain the following data:

$v.type$	satellite or center
$v.degree$	degree of v
$v.adj$	list of adjacent vertices
$v.centers$	list of adjacent centers
$v.inQ$	flag specifying if v being processed

Note that while $v.type$ can be inferred from $v.centers$ and $v.degree$ can be inferred from $v.adj$, it will be convenient to maintain all five pieces of data in the algorithm.

The basic idea behind the online star algorithm is as follows. When a vertex is inserted into (or deleted from) a thresholded similarity graph G_σ, new stars may need to be created and existing stars may need to be destroyed. An existing star is never destroyed unless a satellite is "promoted" to center status. The online star algorithm functions by maintaining a priority queue (indexed by vertex degree), which contains all satellite vertices that have the possibility of being promoted. So long as these enqueued vertices are indeed properly satellites, the existing star cover is correct. The enqueued satellite vertices are processed in order by degree (highest to lowest), with satellite promotion occurring as necessary. Promoting a satellite vertex may destroy one or more existing stars, creating new satellite vertices that have the possibility of being promoted. These satellites are enqueued, and the process repeats. We

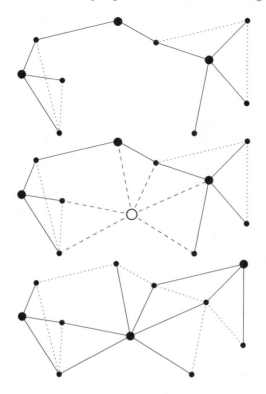

Fig. 5. The star cover change after the insertion of a new vertex. The larger-radius disks denote star centers, the other disks denote satellite vertices. The star edges are denoted by solid lines. The intersatellite edges are denoted by dotted lines. The top figure shows an initial graph and its star cover. The middle figure shows the graph after the insertion of a new document. The bottom figure shows the star cover of the new graph

next describe in some detail the three routines that comprise the online star algorithm.

The INSERT and DELETE procedures are called when a vertex is added to or removed from a thresholded similarity graph, respectively. These procedures appropriately modify the graph structure and initialize the priority queue with all satellite vertices that have the possibility of being promoted. The UPDATE procedure promotes satellites as necessary, destroying existing stars if required, and enqueuing any new satellites that have the possibility of being promoted.

Figure 6 provides the details of the INSERT algorithm. A vertex α with a list of adjacent vertices L is added to a graph G. The priority queue Q is initialized with α (lines 17 and 18) and its adjacent satellite vertices (lines 13 and 14).

```
INSERT(α, L, G_σ)
 1  α.type ← satellite
 2  α.degree ← 0
 3  α.adj ← ∅
 4  α.centers ← ∅
 5  forall β in L
 6     α.degree ← α.degree + 1
 7     β.degree ← β.degree + 1
 8     INSERT(β, α.adj)
 9     INSERT(α, β.adj)
10     if (β.type = center)
11        INSERT(β, α.centers)
12     else
13        β.inQ ← true
14        ENQUEUE(β, Q)
15     endif
16  endfor
17  α.inQ ← true
18  ENQUEUE(α, Q)
19  UPDATE(G_σ)
```

Fig. 6. Pseudocode for INSERT

```
DELETE(α, G_σ)
 1  forall β in α.adj
 2     β.degree ← β.degree − 1
 3     DELETE(α, β.adj)
 4  endfor
 5  if (α.type = satellite)
 6     forall β in α.centers
 7        forall μ in β.adj
 8           if (μ.inQ = false)
 9              μ.inQ ← true
10              ENQUEUE(μ, Q)
11           endif
12        endfor
13     endfor
14  else
15     forall β in α.adj
16        DELETE(α, β.centers)
17        β.inQ ← true
18        ENQUEUE(β, Q)
19     endfor
20  endif
21  UPDATE(G_σ)
```

Fig. 7. Pseudocode for DELETE

The DELETE algorithm presented in Fig. 7 removes vertex α from the graph data structures, and depending on the type of α enqueues its adjacent satellites (lines 15–19) or the satellites of its adjacent centers (lines 6–13).

Finally, the algorithm for UPDATE is shown in Fig. 8. Vertices are organized in a priority queue, and a vertex ϕ of highest degree is processed in each iteration (line 2). The algorithm creates a new star with center ϕ if ϕ has no adjacent centers (lines 3–7) or if all its adjacent centers have lower degree (lines 9–13). The latter case destroys the stars adjacent to ϕ, and their satellites are enqueued (lines 14–23). The cycle is repeated until the queue is empty.

Correctness and Optimizations

The online star cover algorithm is more complex than its offline counterpart. One can show that the online algorithm is correct by proving that it produces the same star cover as the offline algorithm, when the offline algorithm is run on the final graph considered by the online algorithm. We first note, however, that the offline star algorithm need not produce a unique cover. When there are several unmarked vertices of the same highest degree, the algorithm can arbitrarily choose one of them as the next star center. In this context, one can

```
UPDATE(Gσ)
 1  while (Q ≠ ∅)
 2      φ ← EXTRACTMAX(Q)
 3      if (φ.centers = ∅)
 4          φ.type ← center
 5          forall β in φ.adj
 6              INSERT(φ, β.centers)
 7          endfor
 8      else
 9          if (∀δ ∈ φ.centers, δ.degree < φ.degree)
10              φ.type ← center
11              forall β in φ.adj
12                  INSERT(φ, β.centers)
13              endfor
14              forall δ in φ.centers
15                  δ.type ← satellite
16                  forall μ in δ.adj
17                      DELETE(δ, μ.centers)
18                      if (μ.degree ≤ δ.degree ∧ μ.inQ = false)
19                          μ.inQ ← true
20                          ENQUEUE(μ, Q)
21                      endif
22                  endfor
23              endfor
24              φ.centers ← ∅
25          endif
26      endif
27      φ.inQ ← false
28  endwhile
```

Fig. 8. Pseudocode for UPDATE

show that the cover produced by the online star algorithm is the same as one of the covers that can be produced by the offline algorithm. We can view a star cover of G_σ as a correct assignment of types (that is, "center" or "satellite") to the vertices of G_σ. The offline star algorithm assigns correct types to the vertices of G_σ. The online star algorithm is proven correct by induction. The induction invariant is that at all times, the types of all vertices in $V - Q$ are correct, *assuming* that the true type of all vertices in Q is "satellite." This would imply that when Q is empty, all vertices are assigned a correct type, and thus the star cover is correct. Details can be found in [28, 31].

Finally, we note that the online algorithm can be implemented more efficiently than described here. An optimized version of the online algorithm exists, which maintains additional information and uses somewhat different data structures. While the asymptotic running time of the optimized version of the online algorithm is unchanged, the optimized version is often faster in practice. Details can be found in [31].

4.2 Expected Running Time of the Online Algorithm

In this section, we argue that the running time of the online star algorithm is quite efficient, asymptotically matching the running time of the offline star algorithm within logarithmic factors. We first note, however, that there exist worst-case thresholded similarity graphs and corresponding vertex insertion/deletion sequences that cause the online star algorithm to "thrash" (i.e., which cause the entire star cover to change on each inserted or deleted vertex). These graphs and insertion/deletion sequences rarely arise in practice however. An analysis more closely modeling practice is the random graph model [78] in which G_σ is a random graph and the insertion/deletion sequence is random. In this model, the *expected* running time of the online star algorithm can be determined. In the remainder of this section, we argue that the online star algorithm is quite efficient theoretically. In subsequent sections, we provide empirical results that verify this fact for both random data and a large collection of real documents.

The model we use for expected case analysis is the *random graph model* [78]. A random graph $G_{n,p}$ is an undirected graph with n vertices, where each of its possible edges is inserted randomly and independently with probability p. Our problem fits the random graph model if we make the mathematical assumption that "similar" documents are essentially "random perturbations" of one another in the vector space model. This assumption is equivalent to viewing the similarity between two related documents as a random variable. By thresholding the edges of the similarity graph at a fixed value, for each edge of the graph there is a random chance (depending on whether the value of the corresponding random variable is above or below the threshold value) that the edge remains in the graph. This thresholded similarity graph is thus a random graph. While random graphs do not perfectly model the thresholded similarity graphs obtained from actual document corpora (the actual similarity graphs must satisfy various geometric constraints and will be aggregates of many "sets" of "similar" documents), random graphs are easier to analyze, and our experiments provide evidence that theoretical results obtained for random graphs closely match empirical results obtained for thresholded similarity graphs obtained from actual document corpora. As such, we use the random graph model for analysis and experimental verification of the algorithms presented in this chapter (in addition to experiments on actual corpora).

The time required to insert/delete a vertex and its associated edges and to appropriately update the star cover is largely governed by the number of stars that are broken during the update, since breaking stars requires inserting new elements into the priority queue. In practice, very few stars are broken during any given update. This is partly due to the fact that relatively few stars exist at any given time (as compared to the number of vertices or edges in the thresholded similarity graph) and partly to the fact that the likelihood of breaking any individual star is also small. We begin by examining the expected size of a star cover in the random graph model.

Theorem 5. *The expected size of the star cover for $G_{n,p}$ is at most $1 + 2\log n/\log(1/(1-p))$.*

Proof. The star cover algorithm is greedy: it repeatedly selects the unmarked vertex of highest degree as a star center, marking this node and all its adjacent vertices as covered. Each iteration creates a new star. We argue that the number of iterations is at most $1 + 2\log n/\log(1/(1-p))$ for an even weaker algorithm, which merely selects *any* unmarked vertex (at random) to be the next star. The argument relies on the random graph model described earlier.

Consider the (weak) algorithm described earlier which repeatedly selects stars at random from $G_{n,p}$. After i stars have been created, each of the i star centers is marked, and some of the $n - i$ remaining vertices is marked. For any given noncenter vertex, the probability of being adjacent to any given center vertex is p. The probability that a given noncenter vertex remains unmarked is therefore $(1 - p)^i$, and thus its probability of being marked is $1 - (1 - p)^i$. The probability that *all* $n - i$ noncenter vertices are marked is then $\left(1 - (1 - p)^i\right)^{n-i}$. This is the probability that i (random) stars are sufficient to cover $G_{n,p}$. If we let X be a random variable corresponding to the number of star required to cover $G_{n,p}$, we then have

$$\Pr[X \geq i + 1] = 1 - \left(1 - (1 - p)^i\right)^{n-i}.$$

Using the fact that for any discrete random variable Z whose range is $\{1, 2, \ldots, n\}$,

$$\mathrm{E}[Z] = \sum_{i=1}^{n} i \times \Pr[Z = i] = \sum_{i=1}^{n} \Pr[Z \geq i],$$

we then have

$$\mathrm{E}[X] = \sum_{i=0}^{n-1} \left[1 - \left(1 - (1 - p)^i\right)^{n-i}\right].$$

Note that for any $n \geq 1$ and $x \in [0, 1]$, $(1 - x)^n \geq 1 - nx$. We may then derive

$$\mathrm{E}[X] = \sum_{i=0}^{n-1} \left[1 - \left(1 - (1 - p)^i\right)^{n-i}\right]$$

$$\leq \sum_{i=0}^{n-1} \left[1 - \left(1 - (1 - p)^i\right)^{n}\right]$$

$$= \sum_{i=0}^{k-1} \left[1 - \left(1 - (1 - p)^i\right)^{n}\right] + \sum_{i=k}^{n-1} \left[1 - \left(1 - (1 - p)^i\right)^{n}\right]$$

$$\leq \sum_{i=0}^{k-1} 1 + \sum_{i=k}^{n-1} n(1-p)^i$$

$$= k + \sum_{i=k}^{n-1} n(1-p)^i$$

for any k. Selecting k so that $n(1-p)^k = 1/n$ (i.e., $k = 2\log n / \log(1/(1-p)))$, we have

$$\mathrm{E}[X] \leq k + \sum_{i=k}^{n-1} n(1-p)^i$$

$$\leq 2\log n / \log(1/(1-p)) + \sum_{i=k}^{n-1} 1/n$$

$$\leq 2\log n / \log(1/(1-p)) + 1. \qquad \square$$

Combining the above theorem with various facts concerning the behavior of the UPDATE procedure, one can show the following.

Theorem 6. *The expected time required to insert or delete a vertex in a random graph $G_{n,p}$ is $O(np^2 \log^2 n / \log^2(1/(1-p)))$, for any $0 \leq p \leq 1 - \Theta(1)$.*

The proof of this theorem is rather technical; details can be found in [31]. The thresholded similarity graphs obtained in a typical IR setting are almost always dense: there exist many vertices comprising relatively few (but dense) clusters. We obtain dense random graphs when p is a constant. For dense graphs, we have the following corollary.

Corollary 1. *The total expected time to insert n vertices into (an initially empty) dense random graph is $O(n^2 \log^2 n)$.*

Corollary 2. *The total expected time to delete n vertices from (an n vertex) dense random graph is $O(n^2 \log^2 n)$.*

Note that the online insertion result for dense graphs compares favorably to the offline algorithm; both algorithms run in time proportional to the size of the input graph, $\Theta(n^2)$, within logarithmic factors. Empirical results on dense random graphs and actual document collections (detailed in Sect. 4.3) verify this result.

For sparse graphs ($p = \Theta(1/n)$), we note that $1/\ln(1/(1-\epsilon)) \approx 1/\epsilon$ for small ϵ. Thus, the expected time to insert or delete a single vertex is $O(np^2 \log^2 n / \log^2(1/(1-p))) = O(n \log^2 n)$, yielding an asymptotic result identical to that of dense graphs, much larger than what one encounters in practice. This is due to the fact that the number of stars broken (and hence vertices enqueued) is much smaller than the worst-case assumptions assumed in the analysis of the UPDATE procedure. Empirical results on sparse random graphs (detailed in the following section) verify this fact and imply that the total running time of the online insertion algorithm is also proportional to the size of the input graph, $\Theta(n)$, within lower order factors.

4.3 Experimental Validation

To experimentally validate the theoretical results obtained in the random graph model, we conducted efficiency experiments with the online star clustering algorithm using two types of data. The first type of data matches our random graph model and consists of both sparse and dense random graphs. While this type of data is useful as a benchmark for the running time of the algorithm, it does not satisfy the geometric constraints of the vector space model. We also conducted experiments using 2,000 documents from the TREC FBIS collection.

Aggregate Number of Broken Stars

As discussed earlier, the efficiency of the online star algorithm is largely governed by the number of stars that are broken during a vertex insertion or deletion. In our first set of experiments, we examined the aggregate number of broken stars during the insertion of 2,000 vertices into a sparse random graph ($p = 10/n$), a dense random graph ($p = 0.2$), and a graph corresponding to a subset of the TREC FBIS collection thresholded at the mean similarity. The results are given in Fig. 9.

For the sparse random graph, while inserting 2,000 vertices, 2,572 total stars were broken – approximately 1.3 broken stars per vertex insertion on average. For the dense random graph, while inserting 2,000 vertices, 3,973 total stars were broken – approximately 2 broken stars per vertex insertion on average. The thresholded similarity graph corresponding to the TREC FBIS data was much denser, and there were far fewer stars. While inserting 2,000 vertices, 458 total stars were broken – approximately 23 broken stars per 100 vertex insertions on average. Thus, even for moderately large n, the number of broken stars per vertex insertion is a relatively small constant, though we do note the effect of lower order factors especially in the random graph experiments.

Aggregate Running Time

In our second set of experiments, we examined the aggregate running time during the insertion of 2,000 vertices into a sparse random graph ($p = 10/n$), a dense random graph ($p = 0.2$), and a graph corresponding to a subset of the TREC FBIS collection thresholded at the mean similarity. The results are given in Fig. 10.

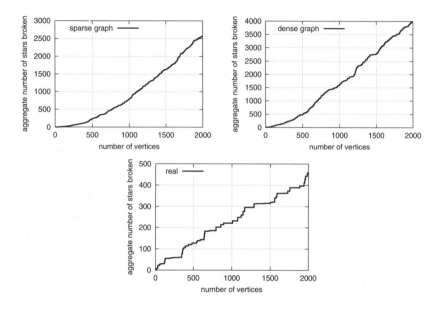

Fig. 9. The dependence of the number of broken stars on the number of inserted vertices in a sparse random graph (top left figure), a dense random graph (top right figure), and the graph corresponding to TREC FBIS data (bottom figure)

Note that for connected input graphs (sparse or dense), the size of the graph is on the order of the number of edges. The experiments depicted in Fig. 10 suggest a running time for the online algorithm, which is linear in the size of the input graph, though lower order factors are presumably present.

5 The Accuracy of Star Clustering

In this section we describe experiments evaluating the performance of the star algorithm with respect to cluster accuracy. We tested the star algorithm against two widely used clustering algorithms in IR: the single link method [429] and the average link method [434]. We used data from the TREC FBIS collection as our testing medium. This TREC collection contains a very large set of documents of which 21,694 have been ascribed relevance judgments with respect to 47 topics. These 21,694 documents were partitioned into 22 separate subcollections of approximately 1,000 documents each for 22 rounds of the following test. For each of the 47 topics, the given collection of documents was clustered with each of the three algorithms, and the cluster that "best" approximated the set of judged relevant documents was returned. To measure the quality of a cluster, we use the standard F measure from IR [429]:

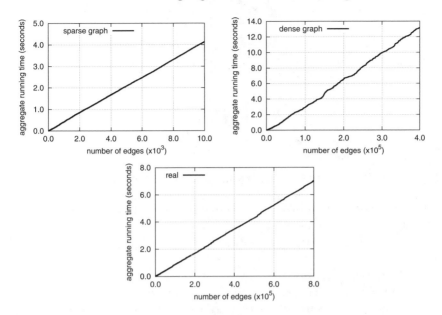

Fig. 10. The dependence of the running time of the online star algorithm on the size of the input graph for a sparse random graph (top left figure), a dense random graph (top right figure), and the graph corresponding to TREC FBIS data (bottom figure)

$$F(p,r) = \frac{2}{(1/p) + (1/r)},$$

where p and r are the *precision* and *recall* of the cluster with respect to the set of documents judged relevant to the topic. Precision is the fraction of returned documents that are correct (i.e., judged relevant), and recall is the fraction of correct documents that are returned. $F(p,r)$ is simply the harmonic mean of the precision and recall; thus, $F(p,r)$ ranges from 0 to 1, where $F(p,r) = 1$ corresponds to perfect precision and recall, and $F(p,r) = 0$ corresponds to either zero precision or zero recall.

For each of the three algorithms, approximately 500 experiments were performed; this is roughly half of the $22 \times 47 = 1,034$ total possible experiments since not all topics were present in all subcollections. In each experiment, the $(p,r,F(p,r))$ values corresponding to the cluster of highest quality were obtained, and these values were averaged over all 500 experiments for each algorithm. The average $(p,r,F(p,r))$ values for the star, average-link, and single-link algorithms were, $(0.77, 0.54, 0.63)$, $(0.83, 0.44, 0.57)$ and $(0.84, 0.41, 0.55)$, respectively. Thus, the star algorithm represents a 10.5% improvement in cluster accuracy with respect to the average-link algorithm and a 14.5% improvement in cluster accuracy with respect to the single-link algorithm.

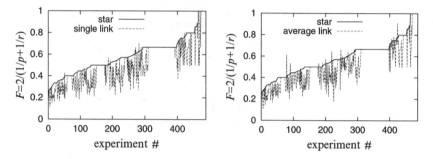

Fig. 11. The F measure for the star clustering algorithm vs. the single link clustering algorithm (left) and the star algorithm vs. the average link algorithm (right). The y axis shows the F measure. The x axis shows the experiment number. Experimental results have been sorted according to the F value for the star algorithm

Figure 11 shows the results of all 500 experiments. The first graph shows the accuracy (F measure) of the star algorithm vs. the single-link algorithm; the second graph shows the accuracy of the star algorithm vs. the average-link algorithm. In each case, the results of the 500 experiments using the star algorithm were sorted according to the F measure (so that the star algorithm results would form a monotonically increasing curve), and the results of both algorithms (star and single-link or star and average-link) were plotted according to this sorted order. While the *average* accuracy of the star algorithm is higher than that of either the single-link or the average-link algorithms, we further note that the star algorithm outperformed each of these algorithms in nearly *every* experiment.

Our experiments show that in general, the star algorithm outperforms single-link by 14.5% and average-link by 10.5%. We repeated this experiment on the same data set, using the entire unpartitioned collection of 21,694 documents, and obtained similar results. The precision, recall, and F values for the star, average-link, and single-link algorithms were $(0.53, 0.32, 0.42)$, $(0.63, 0.25, 0.36)$, and $(0.66, 0.20, 0.30)$, respectively. We note that the F values are worse for all three algorithms on this larger collection and that the star algorithm outperforms the average-link algorithm by 16.7% and the single-link algorithm by 40%. These improvements are significant for IR applications. Given that (1) the star algorithm outperforms the average-link algorithm, (2) it can be used as an online algorithm, (3) it is relatively simple to implement in either of its offline or online forms, and (4) it is efficient, these experiments provide support for using the star algorithm for offline and online information organization.

6 Applications of the Star Clustering Algorithm

We have investigated the use of the star clustering algorithm in a number of different application areas including: (1) automatic information organization systems [26, 27], (2) scalable information organization for large corpora [33], (3) text filtering [29, 30], and (4) persistent queries [32]. In the sections that follow, we briefly highlight this work.

6.1 A System for Information Organization

We have implemented a system for organizing information that uses the star algorithm (see Fig. 12). This organization system consists of an augmented version of the Smart system [18, 378], a user interface we have designed, and an implementation of the star algorithms on top of Smart. To index the documents, we used the Smart search engine with a cosine normalization weighting scheme. We enhanced Smart to compute a document-to-document similarity matrix for a set of retrieved documents or a whole collection. The similarity matrix is used to compute clusters and to visualize the clusters.

The figure shows the interface to the information organization system. The search and organization choices are described at the top. The middle two windows show two views of the organized documents retrieved from the Web or from the database. The left window shows the list of topics, the number of documents in each topic, and a keyword summary for each topic. The right window shows a graphical description of the topics. Each topic corresponds to a disk. The size of the disk is proportional to the number of documents in the topic cluster and the distance between two disks is proportional to the topic similarity between the corresponding topics. The bottom window shows a list of titles for the documents. The three views are connected: a click in one window causes the corresponding information to be highlighted in the other two windows. Double clicking on any cluster (in the right or left middle panes) causes the system to organize and present the documents in that cluster, thus creating a view one level deeper in a hierarchical cluster tree; the "Zoom Out" button allows one to retreat to a higher level in the cluster tree. Details on this system and its variants can be found in [26, 27, 29].

6.2 Scalable Information Organization

The star clustering algorithm implicitly assumes the existence of a thresholded similarity graph. While the running times of the offline and the online star clustering algorithms are linear in the size of the input graph (to within lower order factors), the size of these graphs themselves may be prohibitively large. Consider, for example, an information system containing n documents and a request to organize this system with a relatively low similarity threshold. The resulting graph would in all likelihood be dense, i.e., have $\Omega(n^2)$ edges. If n is large (e.g., millions), just computing the thresholded similarity graph may

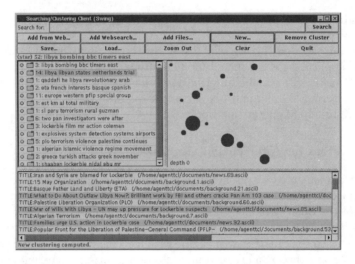

Fig. 12. A system for information organization based on the star clustering algorithm

be prohibitively expensive, let alone running a clustering algorithm on such a graph.

In [33], we propose three methods based on sampling and/or parallelism for generating accurate *approximations* to a star cover in time linear in the number of documents, independent of the size of the thresholded similarity graph.

6.3 Filtering and Persistent Queries

Information filtering and persistent query retrieval are related problems wherein relevant elements of a dynamic stream of documents are sought in order to satisfy a user's information need. The problems differ in how the information need is supplied: in the case of filtering, exemplar documents are supplied by the user, either dynamically or in advance; in the case of persistent query retrieval, a standing query is supplied by the user.

We propose a solution to the problems of information filtering and persistent query retrieval through the use of the star clustering algorithm. The salient features of the systems we propose are (1) the user has access to the *topic structure* of the document collection star clusters; (2) the query (filtering topic) can be formulated as a list of keywords, a set of selected documents, or a set of selected *document clusters*; (3) document filtering is based on *prospective cluster membership*; (4) the user can modify the query by providing relevance feedback on the document clusters and individual documents in the *entire collection*; and (5) the relevant documents adapt as the collection changes. Details can be found in [29, 30, 32].

7 Conclusions

We presented and analyzed an offline clustering algorithm for static information organization and an online clustering algorithm for dynamic information
organization. We described a random graph model for analyzing the running
times of these algorithms, and we showed that in this model, these algorithms
have an expected running time that is linear in the size of the input graph,
to within lower order factors. The data we gathered from experiments with
TREC data lend support for the validity of our model and analyses. Our empirical tests show that both algorithms exhibit linear time performance in the
size of the input graph (to within lower order factors), and that both algorithms produce accurate clusters. In addition, both algorithms are simple and
easy to implement. We believe that efficiency, accuracy, and ease of implementation make these algorithms very practical candidates for use in automatic
information organization systems.

This work departs from previous clustering algorithms often employed in
IR settings, which tend to use a fixed number of clusters for partitioning the
document space. Since the number of clusters produced by our algorithms is
given by the underlying topic structure in the information system, our clusters
are dense and accurate. Our work extends previous results [225] that support
using clustering for browsing applications and presents positive evidence for
the cluster hypothesis. In [26], we argue that by using a clustering algorithm
that guarantees the cluster quality through separation of dissimilar documents
and aggregation of similar documents, clustering is beneficial for information
retrieval tasks that require both high precision and high recall.

Acknowledgments

This research was supported in part by ONR contract N00014-95-1-1204,
DARPA contract F30602-98-2-0107, and NSF grant CCF-0418390.

A Survey of Clustering Data Mining Techniques

P. Berkhin

Summary. *Clustering* is the division of data into groups of similar objects. In clustering, some details are disregarded in exchange for data simplification. Clustering can be viewed as a data modeling technique that provides for concise summaries of the data. Clustering is therefore related to many disciplines and plays an important role in a broad range of applications. The applications of clustering usually deal with large datasets and data with many attributes. Exploration of such data is a subject of data mining. This survey concentrates on clustering algorithms from a data mining perspective.

1 Introduction

We provide a comprehensive review of different clustering techniques in data mining. *Clustering* refers to the division of data into groups of similar objects. Each group, or cluster, consists of objects that are similar to one another and dissimilar to objects in other groups. When representing a quantity of data with a relatively small number of clusters, we achieve some simplification, at the price of some loss of detail (as in lossy data compression, for example). Clustering is a form of data modeling, which puts it in a historical perspective rooted in mathematics and statistics. From a machine learning perspective, clusters correspond to *hidden patterns*, the search for clusters is unsupervised learning, and the resulting system represents a data concept. Therefore, clustering is *unsupervised learning* of a hidden *data concept*. Clustering as applied to data mining applications encounters three additional complications: (a) large databases, (b) objects with many attributes, and (c) attributes of different types. These complications tend to impose severe computational requirements that present real challenges to classic clustering algorithms. These challenges led to the emergence of powerful broadly applicable data mining clustering methods developed on the foundation of classic techniques. These clustering methods are the subject of this survey.

1.1 Notations

To fix the context and clarify terminology, consider a dataset X consisting of data points (which may in turn represent *objects, instances, cases, patterns, tuples, transactions*, and so forth) $x_i = (x_{i1}, \ldots, x_{id})$, $i = 1 : N$, in attribute space A, where each component $x_{il} \in A_l$, $l = 1 : d$, is a numerical or a nominal categorical attribute (which may represent a *feature, variable, dimension, component*, or *field*). For a discussion of attribute data types see [217]. This point-by-attribute data format conceptually corresponds to an $N \times d$ matrix and is used by the majority of algorithms reviewed later. However, data of other formats, such as variable length sequences and heterogeneous data, are not uncommon.

The simplest subset in an attribute space is a direct Cartesian product of subranges $C = \prod C_l \subset A$, $C_l \subset A_l$, called a *segment* (or a *cube, cell*, or *region*). A *unit* is an elementary segment whose subranges consist of a single category value or a small numerical bin. Describing the number of data points per unit represents an extreme case of clustering, a *histogram*. The histogram is a very expensive representation and not a very revealing one. User-driven *segmentation* is another commonly used practice in data exploration that utilizes expert knowledge regarding the importance of certain subdomains. Unlike segmentation, clustering is assumed to be automatic, and so it is unsupervised in the machine learning sense.

The goal of clustering is to assign data points to a finite system of k subsets (clusters). These subsets do not intersect (however, this requirement is sometimes violated in practice), and their union is equal to the full dataset with the possible exception of outliers

$$X = C_1 \bigcup \cdots \bigcup C_k \bigcup C_{outliers}, C_i \bigcap C_j = 0, i \neq j.$$

1.2 Clustering Bibliography at a Glance

General references regarding clustering include [142,155,159,188,218,224,242, 245,265,287,337,405]. A very good introduction to contemporary data mining clustering techniques can be found in Han and Kamber [217].

Clustering is related to many other fields. Clustering has been widely used in statistics [24] and science [328]. The classic introduction to clustering in pattern recognition is given in [143]. For statistical approaches to pattern recognition see [126] and [180]. Machine learning clustering algorithms were applied to image segmentation and computer vision [243]. Clustering can be viewed as a density estimation problem. This is the subject of traditional multivariate statistical estimation [391]. Clustering is also widely used for data compression in image processing, which is also known as vector quantization [185]. Data fitting in numerical analysis provides still another venue in data modeling [121].

This survey's emphasis is on clustering in data mining. Such clustering is characterized by large datasets with many attributes of different types. Though we do not even try to review particular applications, many important ideas are related to specific fields. We briefly mention:

- Information retrieval and text mining [121, 129, 407];
- Spatial database applications, dealing with GIS or astronomical data, for example [151, 383, 446];
- Sequence and heterogeneous data analysis [95];
- Web applications [113, 168, 226];
- DNA analysis in computational biology [55].

These and many other application-specific developments are beyond our scope, but some general techniques have been been applied widely. These techniques and classic clustering algorithms related to them are surveyed below.

1.3 Plan of Further Presentation

Classification of clustering algorithms is neither straightforward nor canonical. In fact, the different classes of algorithms overlap. Traditional clustering techniques are broadly divided into *hierarchical* and *partitioning*. Hierarchical clustering is further subdivided into *agglomerative* and *divisive*. The basics of hierarchical clustering include the Lance–Williams formula, the idea of conceptual clustering, the now classic algorithms SLINK and COBWEB, as well as the newer algorithms CURE and CHAMELEON. We survey these algorithms in Sect. 2.

While hierarchical algorithms gradually (dis)assemble points into clusters (as crystals grow), partitioning algorithms learn clusters directly. In doing so, they try to discover clusters either by iteratively relocating points between subsets or by identifying areas heavily populated with data.

Algorithms of the first kind are called *Partitioning Relocation Clustering*. They are further classified into *probabilistic* clustering (EM framework, algorithms SNOB, AUTOCLASS, MCLUST), k-medoids methods (algorithms PAM, CLARA, CLARANS, and their extensions), and the various k-means methods. They are presented in Sect. 3. Such methods concentrate on how well points fit into their clusters and tend to build clusters of proper convex shapes.

Partitioning algorithms of the second type are surveyed in Sect. 4. These algorithms attempt to discover dense connected components of data, which are flexible in terms of their shape. Density-based connectivity is used in the algorithms DBSCAN, OPTICS, and DBCLASD, while the algorithm DENCLUE exploits space density functions. These algorithms are less sensitive to outliers and can discover clusters of irregular shape. They usually work with low-dimensional numerical data, known as *spatial* data. Spatial objects may include points, but also geometrically extended objects (as in the algorithm GDBSCAN).

Some algorithms work with data indirectly by constructing summaries of data over the attribute space subsets. These algorithms perform space segmentation and then aggregate appropriate segments. We discuss these algorithms in Sect. 5. These algorithms frequently use hierarchical agglomeration as one phase of processing. Algorithms BANG, STING, WaveCluster, and FC are discussed in this section. Grid-based methods are fast and handle outliers well. The grid-based methodology is also used as an intermediate step in many other algorithms (for example, CLIQUE and MAFIA).

Categorical data are intimately connected with transactional databases. The concept of similarity alone is not sufficient for clustering such data. The idea of categorical data co-occurrence comes to the rescue. The algorithms ROCK, SNN, and CACTUS are surveyed in Sect. 6. Clustering of categorical data grows more difficult as the number of items involved increases. To help with this problem, the effort is shifted from data clustering to preclustering of items or categorical attribute values. Developments based on *hypergraph* partitioning and the algorithm STIRR exemplify this approach.

Many other clustering techniques have been developed, primarily in machine learning, that either have theoretical significance, are used traditionally outside the data mining community, or do not fit in previously outlined categories. The boundary is blurred. In Sect. 7 we discuss the emerging direction of *constraint-based clustering*, the important research field of *graph partitioning*, and the relationship of clustering to *supervised learning*, *gradient descent*, *artificial neural networks*, and *evolutionary methods*.

Data mining primarily works with large databases. Clustering large datasets presents scalability problems reviewed in Sect. 8. We discuss algorithms like DIGNET, BIRCH and other data squashing techniques, and Hoeffding or Chernoff bounds.

Another trait of real-life data is high dimensionality. Corresponding developments are surveyed in Sect. 9. The trouble with high dimensionality comes from a decrease in metric separation as the dimension grows. One approach to *dimensionality reduction* uses attribute transformations (e.g., DFT, PCA, wavelets). Another way to address the problem is through *subspace clustering* (as in algorithms CLIQUE, MAFIA, ENCLUS, OPTIGRID, PROCLUS, ORCLUS). Still another approach clusters attributes in groups and uses their derived proxies to cluster objects. This double clustering is known as *coclustering*.

Issues common to different clustering methods are overviewed in Sect. 10. We discuss *assessment* of results, determination of the appropriate number of clusters to build, *data preprocessing*, *proximity measures*, and handling of *outliers*.

For the reader's convenience we provide a *classification of clustering algorithms* closely followed by this survey:

- Hierarchical methods
    Agglomerative algorithms

    Divisive algorithms
- Partitioning relocation methods
    Probabilistic clustering
    k-medoids methods
    k-means methods
- Density-based partitioning methods
    Density-based connectivity clustering
    Density functions clustering
- Grid-based methods
- Methods based on co-occurrence of categorical data
- Other clustering techniques
    Constraint-based clustering
    Graph partitioning
    Clustering algorithms and supervised learning
    Clustering algorithms in machine learning
- Scalable clustering algorithms
- Algorithms for high-dimensional data
    Subspace clustering
    Coclustering techniques

1.4 Important Issues

The properties of clustering algorithms of concern in data mining include:

- Type of attributes an algorithm can handle
- Scalability to large datasets
- Ability to work with high-dimensional data
- Ability to find clusters of irregular shape
- Handling outliers
- Time complexity (we often simply use the term *complexity*)
- Data order dependency
- Labeling or assignment (hard or strict vs. soft or fuzzy)
- Reliance on a priori knowledge and user-defined parameters
- Interpretability of results

Realistically, with every algorithm we discuss only some of these properties. This list is not intended to be exhaustive. For example, as appropriate, we also discuss the algorithm's ability to work in a predefined memory buffer, to restart, and to provide intermediate solutions.

2 Hierarchical Clustering

Hierarchical clustering combines data objects into clusters, those clusters into larger clusters, and so forth, creating a hierarchy. A tree representing this hierarchy of clusters is known as a *dendrogram*. Individual data objects are the leaves of the tree, and the interior nodes are nonempty clusters. Sibling nodes partition the points covered by their common parent. This allows exploring data at different levels of granularity. Hierarchical clustering methods are categorized into *agglomerative* (bottom-up) and *divisive* (top-down) [242, 265] approaches. An agglomerative clustering starts with one-point (singleton) clusters and recursively merges two or more of the most similar clusters. A divisive clustering starts with a single cluster containing all data points and recursively splits that cluster into appropriate subclusters. The process continues until a stopping criterion (frequently, the requested number k of clusters) is achieved. The advantages of hierarchical clustering include:

- Flexibility regarding the level of granularity,
- Ease of handling any form of similarity or distance,
- Applicability to any attribute type.

The disadvantages of hierarchical clustering are:

- The difficulty of choosing the right stopping criteria,
- Most hierarchical algorithms do not revisit (intermediate) clusters once they are constructed.

The classic approaches to hierarchical clustering are presented in Sect. 2.1. Hierarchical clustering based on linkage metrics results in clusters of proper (convex) shapes. Active contemporary efforts to build cluster systems that incorporate our intuitive concept of clusters as connected components of arbitrary shape, including the algorithms CURE and CHAMELEON, are surveyed in Sect. 2.2. Divisive techniques based on binary taxonomies are presented in Sect. 2.3. Section 7.6 contains information related to incremental learning, model-based clustering, and cluster refinement.

2.1 Linkage Metrics

In hierarchical clustering, our regular point-by-attribute data representation is often of secondary importance. Instead, hierarchical clustering deals with the $N \times N$ matrix of distances (dissimilarities) or similarities between training points sometimes called a *connectivity* matrix. The so-called linkage metrics are constructed from elements of this matrix. For a large data set, keeping a connectivity matrix in memory is impractical. Instead, different techniques are used to *sparsify* (introduce zeros into) the connectivity matrix. This can be done by omitting entries smaller than a certain threshold, by using only a certain subset of data representatives, or by keeping with each point only a certain number of its nearest neighbors (for nearest neighbor chains see [353]). The way we process the original (dis)similarity matrix and construct a linkage metric reflects our a priori ideas about the data model.

With the (sparsified) connectivity matrix we can associate the weighted connectivity graph $G(X, E)$ whose vertices X are data points, and edges E and their weights are defined by the connectivity matrix. This establishes a connection between hierarchical clustering and graph partitioning. One of the most striking developments in hierarchical clustering is the BIRCH algorithm, discussed in Sect. 8.

Hierarchical clustering initializes a cluster system as a set of singleton clusters (agglomerative case) or a single cluster of all points (divisive case) and proceeds iteratively merging or splitting the most appropriate cluster(s) until the stopping criterion is satisfied. The appropriateness of a cluster(s) for merging or splitting depends on the (dis)similarity of cluster(s) elements. This reflects a general presumption that clusters consist of similar points. An important example of dissimilarity between two points is the distance between them.

To merge or split subsets of points rather than individual points, the distance between individual points has to be generalized to the distance between subsets. Such a derived proximity measure is called a *linkage metric*. The type of linkage metric used has a significant impact on hierarchical algorithms, because it reflects a particular concept of *closeness* and *connectivity*. Important intercluster linkage metrics [346, 353] include *single link*, *average link*, and *complete link*. The underlying dissimilarity measure (usually, distance) is computed for every pair of nodes with one node in the first set and another node in the second set. A specific operation such as minimum (single link), average (average link), or maximum (complete link) is applied to pairwise dissimilarity measures:

$$d(C_1, C_2) = Op\left\{d(x, y), x \in C_1, y \in C_2\right\}.$$

Early examples include the algorithm SLINK [396], which implements single link ($Op = \min$), Voorhees' method [433], which implements average link ($Op = \text{Avr}$), and the algorithm CLINK [125], which implements complete link ($Op = \max$). SLINK, for example, is related to the problem of finding the Euclidean minimal spanning tree [449] and has $O(N^2)$ complexity. The methods using intercluster distances defined in terms of pairs of nodes (one in each respective cluster) are naturally related to the connectivity graph $G(X, E)$ introduced earlier, because every data partition corresponds to a graph partition. Such methods can be augmented by the so-called *geometric* methods in which a cluster is represented by its central point. Assuming numerical attributes, the center point is defined as a *centroid* or an average of two cluster centroids subject to agglomeration, resulting in centroid, median, and minimum variance linkage metrics.

All of the above linkage metrics can be derived from the Lance–Williams updating formula [301]:

$$d(C_i \bigcup C_j, C_k) = a(i)d(C_i, C_k) + a(j)d(C_j, C_k) + b \cdot d(C_i, C_j) \\ + c\left|d(C_i, C_k) - d(C_j, C_k)\right|.$$

Here a, b, and c are coefficients corresponding to a particular linkage. This Lance–Williams formula expresses a linkage metric between a union of the two clusters and the third cluster in terms of underlying nodes, and it is crucial to making the dis(similarity) computations feasible. Surveys of linkage metrics can be found in [123, 345]. When distance is used as a base measure, linkage metrics capture intercluster proximity. However, a similarity-based view that results in intracluster connectivity considerations is also used, for example, in the original average link agglomeration (Group-Average Method) [242].

Under reasonable assumptions, such as the reducibility condition, which graph methods satisfy, linkage metrics methods have $O\left(N^2\right)$ time complexity [353]. Despite the unfavorable time complexity, these algorithms are widely used. As an example, the algorithm AGNES (AGlomerative NESting) [265] is used in S-Plus.

When the connectivity $N \times N$ matrix is sparsified, graph methods directly dealing with the connectivity graph G can be used. In particular, the hierarchical divisive MST (Minimum Spanning Tree) algorithm is based on graph partitioning [242].

2.2 Hierarchical Clusters of Arbitrary Shapes

For spatial data, linkage metrics based on Euclidean distance naturally generate clusters of convex shapes. Meanwhile, visual inspection of spatial images frequently reveals clusters with more complex shapes.

Guha et al. [207] introduced the hierarchical agglomerative clustering algorithm CURE (Clustering Using REpresentatives). This algorithm has a number of novel and important features. CURE takes special steps to handle outliers and to provide labeling in the assignment stage. It also uses two techniques to achieve scalability: data sampling (Sect. 8), and data partitioning. CURE creates p partitions, so that fine granularity clusters are constructed in partitions first. A major feature of CURE is that it represents a cluster by a fixed number, c, of points scattered around it. The distance between two clusters used in the agglomerative process is the minimum of distances between two scattered representatives. Therefore, CURE takes a middle approach between the graph (all-points) methods and the geometric (one centroid) methods. Single link and average link closeness are replaced by representatives' aggregate closeness. Selecting representatives scattered around a cluster makes it possible to cover nonspherical shapes. As before, agglomeration continues until the requested number k of clusters is achieved. CURE employs one additional trick: originally selected scattered points are shrunk to the geometric centroid of the cluster by a user-specified factor α. Shrinkage decreases the impact of outliers; outliers happen to be located further from the cluster centroid than the other scattered representatives. CURE is capable of finding clusters of different shapes and sizes. Because CURE uses sampling, estimation of its complexity is not straightforward. For low-dimensional data, Guha et al. provide a complexity estimate of $O(N_{\text{sample}}^2)$ defined in terms of the sample size. More exact bounds depend on the input parameters, which include the shrink factor α, the number of representative points c, the number of partitions p, as well as the sample size. Figure 1a illustrates agglomeration in CURE. Three clusters, each with three representatives, are shown before and after the merge and shrinkage. The two closest representatives are connected.

While the CURE algorithm works with numerical attributes (particularly low-dimensional spatial data), the algorithm ROCK developed by the same researchers [208] targets hierarchical agglomerative clustering for categorical attributes. ROCK is discussed in Sect. 6.

The hierarchical agglomerative algorithm CHAMELEON developed by Karypis et al. [260] uses the connectivity graph G corresponding to the K-nearest neighbor model sparsification of the connectivity matrix: the edges of K most similar points to any given point are preserved, and the rest are pruned. CHAMELEON has two stages. In the first stage, small tight clusters are built, which are input to the second stage. This involves a graph partitioning [460]. In the second stage, an agglomerative process is performed, in which measures of relative interconnectivity $RI(C_i, C_j)$ and relative closeness $RC(C_i, C_j)$ are used. Both measures are locally normalized by internal interconnectivity and closeness of clusters C_i and C_j. In this sense the modeling is *dynamic* and depends on data locally. Normalization involves certain nonobvious graph operations [460]. CHAMELEON relies on graph partitioning implemented in the library HMETIS (as discussed in Sect. 6). Agglomerative process depends on user-provided thresholds. A decision to merge is made based on the combination

$$RI(C_i, C_j) \cdot RC(C_i, C_j)^{\alpha}$$

of local measures. The CHAMELEON algorithm does not depend on assumptions about the data model and has been shown to find clusters of different shapes, densities, and sizes in 2D (two-dimensional) space. CHAMELEON has complexity $O(Nm + N\log(N) + m^2\log(m))$, where m is the number of

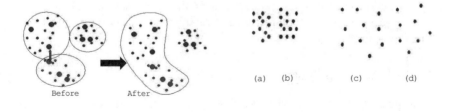

(a) Algorithm CURE   (b) Algorithm CHAMELEON

Fig. 1. Agglomeration in clusters of arbitrary shapes

subclusters built during the first initialization phase. Figure 1b (analogous to the one in [260]) clarifies the difference between CHAMELEON and CURE. It presents a choice of four clusters (a)–(d) for a merge. While CURE would merge clusters (a) and (b), CHAMELEON makes the intuitively better choice of merging (c) and (d).

2.3 Binary Divisive Partitioning

Binary taxonomies are useful in linguistics, information retrieval, and document clustering applications. Linear algebra methods, such as those based on the singular value decomposition (SVD), are used in collaborative filtering and information retrieval [60]. Application of the SVD to hierarchical divisive clustering of document collections resulted in the PDDP (Principal Direction Divisive Partitioning) algorithm by Boley [76]. In our notation, object x is a document, its lth attribute corresponds to a word (*index term*), and a matrix X entry x_{il} is a measure (e.g., TF-IDF) of l-term frequency in a document x. PDDP begins with the SVD of the matrix

$$(X - e\bar{x}), \bar{x} = \frac{1}{N} \sum_{i=1:N} x_i, e = (1, \dots, l)^{\mathrm{T}}.$$

PDDP bisects data in Euclidean space by a hyperplane that passes through the data centroid orthogonal to the eigenvector with the largest singular value. A k-way split is also possible if the k largest singular values are considered. Bisecting is a good way to categorize documents if the goal is to create a binary tree. When k-means (2-means) is used for bisecting, the dividing hyperplane is orthogonal to the line connecting the two centroids. The comparative study of SVD vs. k-means approaches [385] can be consulted for further references. Hierarchical divisive bisecting k-means was proven [407] to be preferable to PDDP for document clustering.

While PDDP or 2-means are concerned with how to split a cluster, the problem of which cluster to split is also important. Simple strategies are: (1) split each node at a given level, (2) split the cluster with highest cardinality, and (3) split the cluster with the largest intracluster variance. All three strategies have problems. For a more detailed analysis of this subject and better strategies, see [386].

2.4 Other Developments

One of the early agglomerative clustering algorithms, Ward's method [441], is based not on a linkage metric, but on an objective function of the sort used in k-means. The merger decision is viewed in terms of its effect on the objective function.

The popular hierarchical clustering algorithm for categorical data COB-WEB [164] has two important qualities. First, it utilizes *incremental* learning.

Instead of following divisive or agglomerative approaches, COBWEB dynamically builds a dendrogram by processing one data point at a time. Second, COBWEB is an example of *conceptual* or *model-based* learning. This means that each cluster is considered as a model that can be described intrinsically, rather than as a collection of points assigned to it. COBWEB's dendrogram is therefore an example of what are called *classification trees*. Each tree node (cluster) C is associated with the conditional probabilities for categorical attribute-values pairs,

$$Pr(x_l = \nu_{lp} | C), \; l = 1 : d, \; p = 1 : |A_l| \,.$$

This can be recognized as a C-specific Naïve Bayes classifier. During the construction of the classification tree, every new point is passed down the tree and the tree is potentially updated (by an insert/split/merge/create operation). Decisions are based on the *category utility* function [114]

$$CU \{C_1, \ldots, C_k\} = \frac{1}{k} \left(\sum_{j=1:k} CU(C_j) \right)$$

$$CU(C_j) = \sum_{l,p} \left((Pr(x_l = \nu_{lp} | C_j)^2 - (Pr(x_l = \nu_{lp})^2 \right) \,.$$

Category utility is similar to the GINI index, in that it rewards clusters C_j for increases in predictability of the categorical attribute values ν_{lp}. Being incremental, COBWEB is fast with a complexity of $O(tN)$, though it depends nonlinearly on tree characteristics packed into a constant t. There is a similar incremental hierarchical algorithm for all numerical attributes called CLASSIT [184]. CLASSIT associates normal distributions with cluster nodes. Both algorithms can result in highly unbalanced trees.

Chiu et al. [111] proposed another *conceptual* or *model-based* approach to hierarchical clustering. This development contains several useful features, such as the extension of scalability preprocessing to categorical attributes, outlier handling, and a two-step strategy for monitoring the number of clusters including BIC (defined later). A model associated with a cluster covers both numerical and categorical attributes and constitutes a blend of Gaussian and multinomial models. Denote corresponding multivariate parameters by θ. With every cluster C we associate a logarithm of its (classification) likelihood

$$l_C = \sum_{x_i \in C} \log \left(p \left(x_i | \theta \right) \right) \,.$$

The algorithm uses *maximum likelihood estimates* for parameter θ. The distance between two clusters is defined (instead of a linkage metric) as a decrease in log-likelihood,

$$d(C_1, C_2) = l_{C_1} + l_{C_2} - l_{C_1 \cup C_2},$$

caused by merging the two clusters under consideration. The agglomerative process continues until the stopping criterion is satisfied. As such, determination of the best k is automatic. This algorithm was used in the commercial implementation of SPSS Clementine. The complexity of the algorithm is linear in N for the summarization phase.

In traditional hierarchical clustering, once a point is assigned to a cluster, the assignment is not changed due to its greedy approach: after a merge or a split decision is made, the decision is not reconsidered. Though COBWEB does reconsider its decisions, its improvement strategy is so primitive that the resulting classification tree can also have subpar quality (though it runs fast). Fisher [165] studied iterative hierarchical cluster redistribution to improve the clustering in a given dendrogram. Karypis et al. [261] also researched refinement for hierarchical clustering. In particular, they brought attention to a relation of such a refinement to a well-studied refinement of k-way graph partitioning [270]. For a review of parallel implementations of hierarchical clustering, see [353].

3 Partitioning Relocation Clustering

In this section, we survey data partitioning algorithms that divide data into several subsets. Because checking all possible subset systems is computationally infeasible, certain greedy heuristics are referred to collectively as *iterative optimization*. Iterative optimization refers to different *relocation* schemes that iteratively reassign points between the k clusters. Unlike traditional hierarchical methods, in which clusters are not revisited after being constructed, relocation algorithms can gradually improve clusters. With appropriate data, this results in high quality clusters.

One approach to data partitioning is to take a *conceptual* point of view that identifies a cluster with a certain model whose unknown parameters have to be found. More specifically, *probabilistic* models assume that the data come from a mixture of several populations whose distributions we wish to characterize. Corresponding algorithms are described in Sect. 3.1. One advantage of probabilistic methods is the interpretability of the constructed clusters. Having concise cluster representations also allows inexpensive computation of intracluster measures of fit that give rise to a global *objective function* (see *log-likelihood* in Sect. 3.1).

Another approach starts with the definition of an objective function depending on a partition. As we have seen (in Sect. 2.1), pairwise distances or similarities can be used to compute measures of inter- and intracluster relations. In iterative improvement approaches such pairwise computations would be too expensive. For this reason, in the k-means and k-medoids approaches, each cluster is associated with a unique cluster representative. Now the computation of an objective function becomes linear in N (and in the number of clusters $k \ll N$). The difference between these k-means and

k-medoid partitioning relocation algorithms is related to how representatives are constructed. Each k-medoid is one of the points. Representation by k-medoids has two advantages: it presents no limitations on attribute types and the choice of medoids is dictated by the location of a predominant fraction of points inside a cluster and, therefore, it is insensitive to the presence of outliers. In k-means a cluster is represented by its centroid, which is a mean (usually weighted average) of points within a cluster. This works conveniently only with numerical attributes and can be negatively affected by a single outlier. On the other hand, centroids have the advantage of clear geometric and statistical meaning. The corresponding algorithms are reviewed in Sects. 3.2 and 3.3.

3.1 Probabilistic Clustering

In the *probabilistic* approach, data are considered to be a sample independently drawn from a *mixture model* of several probability distributions [331]. We assume that data points are generated by: (a) randomly picking a model j with probability $\tau_j, j = 1 : k$, and (b) drawing a point x from a corresponding distribution. The area around the mean of each (supposedly unimodal) distribution constitutes a natural cluster. So we associate a cluster with a corresponding distribution's parameters such as mean, variance, etc. Each data point carries not only its (observable) attributes, but also a (hidden) cluster ID (*class* in pattern recognition). A point x is assumed to belong to one and only one cluster(model) with the probabilities $\Pr(C_j|x)$ that we try to *estimate*. The overall *likelihood* of the training data is its probability of being drawn from a given mixture model

$$\Pr(X|C) = \prod_{i=1:N} \sum_{j=1:k} \tau_j \Pr(x_i|C_j).$$

Log-likelihood $\log(L(X|C))$ serves as an objective function, which gives rise to the *Expectation-Maximization* (EM) method. For a quick introduction to EM, see [340]. Detailed descriptions and numerous references regarding this topic can be found in [126] and [332]. EM is a two-step iterative optimization. Step (E) estimates probabilities $\Pr(x|C_j)$, which is equivalent to a soft (fuzzy) reassignment. Step (M) finds an approximation to the mixture model, given the current soft assignments. This amounts to finding the mixture model parameters that maximize the log-likelihood. The process continues until the log-likelihood convergences.

Restarting and other tricks are used to facilitate finding better local optima. Moore [343] suggested an acceleration of the EM method based on a special data index, *KD* tree. Data are divided at each node into two descendants by splitting the widest attribute at the center of its range. Each node stores sufficient statistics (including the covariance matrix), to allow reconsideration of point assignment decisions (see Sect. 8). Approximate computing over a pruned tree accelerates EM iterations.

Probabilistic clustering has some important features:

- It can be modified to handle points that are recodes of complex structure,
- It can be stopped and resumed with consecutive batches of data, because clusters have representation totally independent from sets of points,
- At any stage of the iterative process the intermediate mixture model can be used to assign points to clusters (online property),
- It results in easily interpretable cluster systems.

Because the mixture model has a clear probabilistic foundation, the determination of the most suitable number of clusters k becomes more tractable. From a data mining perspective, excessive parameter setting can cause overfitting, while from a probabilistic perspective, the number of parameters can be addressed within the Bayesian framework.

The algorithm SNOB [435] uses a mixture model in conjunction with the MML principle (regarding terms MML and BIC see Sect. 10.2). Cheeseman and Stutz [105] developed the algorithm AUTOCLASS that utilizes a mixture model and covers a broad variety of distributions, including Bernoulli, Poisson, Gaussian, and log-normal distributions. Beyond fitting a particular fixed mixture model, AUTOCLASS extends the search to different models and different values of k. To do this AUTOCLASS relies heavily on Bayesian methodology, in which a model's complexity is reflected through certain coefficients (priors) in the expression for the likelihood previously dependent only on parameter values. This algorithm has a history of industrial usage. Finally, the algorithm MCLUST [172] is a software package (commercially linked with S-PLUS) for hierarchical, mixture model clustering, and discriminant analysis using BIC for estimation of goodness of fit. MCLUST uses Gaussian models with ellipsoids of different volumes, shapes, and orientations.

An important property of probabilistic clustering is that the mixture model can be naturally generalized to cluster *heterogeneous* data. This is important in practice when a data object corresponding to an individual person, for example, has both multivariate static data (demographics) in combination with variable length dynamic data (customer profile) [403]. The dynamic data can consist of finite sequences subject to a first-order Markov model with a transition matrix dependent on a cluster. This framework also covers data objects consisting of *several* sequences, where the number n_i of sequences per object x_i is subject to a geometric distribution [94]. To emulate Web browsing sessions of different lengths, for example, a finite-state Markov model (in this case transitional probabilities between Web site pages) has to be augmented with a special "end" state. Cadez et al. [95] used this mixture model for customer profiling based on transactional information.

Model-based clustering is also used in a hierarchical framework: COB-WEB, CLASSIT, and developments in Chiu et al. [111] have already been presented earlier. Another early example of conceptual clustering is algorithm CLUSTER/2 [335].

3.2 k-Medoids Methods

In k-medoids methods a cluster is represented by one of its points. We have already mentioned that this is an easy solution because it covers any attribute type and medoids are insensitive to outliers because peripheral cluster points do not affect them. When medoids are selected, clusters are defined as subsets of points close to respective medoids, and the objective function is defined as the averaged distance or another dissimilarity measure between a point and the corresponding medoid.

Two early versions of k-medoid methods are the algorithms PAM (Partitioning Around Medoids) and CLARA (Clustering LARge Applications) [265]. PAM uses an iterative optimization that combines relocation of points between perspective clusters with renominating the points as potential medoids. The guiding principle for the process is to monitor the effect on an objective function, which, obviously, is a costly strategy. CLARA uses several (five) samples, each with $40{+}2k$ points, which are each subjected to PAM. The whole dataset is assigned to resulting medoids, the objective function is computed, and the best system of medoids is retained.

Further progress is associated with Ng and Han [349] who introduced the algorithm CLARANS (Clustering Large Applications based upon RANdomized Search) in the context of clustering in spatial databases. They considered a graph whose nodes are the sets of k medoids and an edge connects two nodes if these nodes differ by exactly one medoid. While CLARA compares very few neighbors corresponding to a fixed small sample, CLARANS uses random search to generate neighbors by starting with an arbitrary node and randomly checking *maxneighbor* neighbors. If a neighbor represents a better partition, the process continues with this new node. Otherwise, a local minimum is found, and the algorithm restarts until *numlocal* local minima are found (value *numlocal*=2 is recommended). The best node (set of medoids) is returned for the formation of a resulting partition. The complexity of CLARANS is $O\left(N^2\right)$ in terms of the number of points. Ester et al. [153] extended CLARANS to spatial VLDB. They used R^* trees [51] to relax the original requirement that all the data reside in core memory: at any given moment data exploration is *focused* on a branch of the whole data tree.

3.3 k-Means Methods

The k-means algorithm [223, 224] is by far the most popular clustering tool used nowadays in scientific and industrial applications. The name comes from representing each of the k clusters C_j by the mean (or weighted average) c_j of its points, the so-called *centroid*. While this representation does not work well with categorical attributes, it makes good sense from a geometrical and statistical perspective for numerical attributes. The sum of distances between elements of a set of points and its centroid expressed through an appropriate distance function is used as the objective function. For example, the L_2

norm-based objective function, the sum of the squares of errors between the points and the corresponding centroids, is equal to the total intracluster variance

$$E(C) = \sum_{j=1:k} \sum_{x_i \in C_j} \|x_i - c_j\|^2 .$$

The sum of the squares of errors (SSE) can be regarded as the negative of the log-likelihood for a normally distributed mixture model and is widely used in statistics. Therefore, the k-means algorithm can be derived from a general probabilistic framework (see Sect. 3.1) [340]. Note that only means are estimated. A simple modification would normalize individual errors by cluster radii (cluster standard deviation), which makes a lot of sense when clusters have different dispersions. An objective function based on the L_2 norm has many unique algebraic properties. For example, it coincides with pairwise errors,

$$E'(C) = \frac{1}{2} \sum_{j=1:k} \sum_{x_i, y_i \in C_j} \|x_i - y_i\|^2 ,$$

and with the difference between the total data variance and the intercluster variance. Therefore, cluster separation and cluster tightness are achieved simultaneously.

Two versions of k-means iterative optimization are known. The first version is similar to the EM algorithm and consists of two-step major iterations that: (1) reassign all the points to their nearest centroids, and (2) recompute centroids of newly assembled groups. Iterations continue until a stopping criterion is achieved (for example, no reassignments happen). This version, known as Forgy's algorithm [166], has many advantages:

- It easily works with any L_p norm,
- It allows straightforward parallelization [135]
- It does not depend on to data ordering.

The second (classic in iterative optimization) version of k-means reassigns points based on a detailed analysis of how moving a point from its current cluster to any other cluster would affect the objective function. If a move has a positive effect, the point is relocated and the two centroids are recomputed. It is not clear that this version is computationally feasible, because the outlined analysis requires an inner loop over all member points of involved clusters affected by centroids shifts. However, in the L_2 case it is known from [58,143] that computing the impact on a potential cluster can be algebraically reduced to finding a single distance from its centroid to a point in question. Therefore, in this case both versions have the same computational complexity.

There is experimental evidence that compared with Forgy's algorithm, the second (classic) version frequently yields better results [303,407]. In particular, Dhillon et al. [132] noticed that a Forgy's spherical k-means (using cosine similarity instead of Euclidean distance) has a tendency to get stuck when applied to document collections. They noticed that a version that reassigned

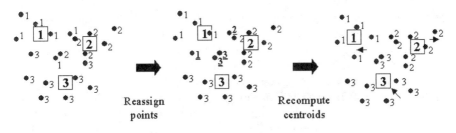

Two-step major iterations (Forgy's algorithm)

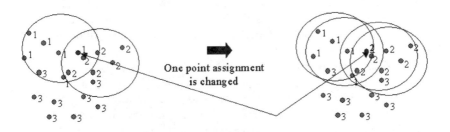

Iterative optimization (with centroid recomputation)

Fig. 2. k-Means algorithm

points and immediately recomputed centroids works much better. Figure 2 illustrates both implementations.

Besides these two versions, there have been other attempts to find better k-means objective functions. For example, the early algorithm ISODATA [42] used merges and splits of intermediate clusters.

The popularity of the k-means algorithm is well deserved, since it is easily understood, easily implemented, and based on the firm foundation of analysis of variances. The k-means algorithm also has certain shortcomings:

- The result depends greatly on the initial guess of centroids,
- The computed local optimum may be quite different from the global one,
- It is not obvious how to choose a good value for k,
- The process is sensitive to outliers,
- The basic algorithm is not scalable,
- Only numerical attributes are covered,
- Resulting clusters can be unbalanced (in Forgy's version, even empty).

A simple way to mitigate the affects of cluster initialization was suggested by Bradley and Fayyad [84]. First, k-means is performed on several small samples of data with a random initial guess. Each of these constructed systems is then used as a potential initialization for a union of all the samples.

Centroids of the best system constructed this way are suggested as an intelligent initial guess to ignite the k-means algorithm on the full data. Another interesting attempt [36] is based on genetic algorithms, as discussed later. No initialization actually guarantees a global minimum for k-means. This is a general problem in combinatorial optimization, which is usually tackled by allowing uphill movements. In our context, simulated annealing was suggested in [91]. Zhang [457] suggested another way to rectify the optimization process by soft assignment of points to different clusters with appropriate weights (as EM does), rather than moving them decisively from one cluster to another. The weights take into account how well a point fits into the recipient cluster. This process involves the so-called *harmonic means*. In this regard, we wish to clarify that the EM algorithm makes soft (fractional) assignments, while the reassignment step in Forgy's version exercises "winner-take-all" or hard assignment. A brilliant earlier analysis of where this subtle difference leads has been conducted by Kearns et al. [266].

For a thorough treatment of k-means scalability, see Bradley et al.'s excellent study [85] (also see Sect. 8 for a general discussion). A generic method to achieve scalability is to preprocess or *squash* the data. Such preprocessing usually also takes care of outliers. Preprocessing has drawbacks. It results in approximations that sometimes negatively affect final cluster quality. Pelleg and Moore [357] suggested how to directly (without any squashing) accelerate the k-means iterative process by utilizing KD trees [343]. The algorithm X-means [358] goes a step further: in addition to accelerating the iterative process, it tries to incorporate a search for the best k in the process itself. While more comprehensive criteria for finding optimal k require running independent k-means and then comparing the results (costly experimentation), X-means tries to split a part of the already constructed cluster based on the outcome of the BIC criterion. This gives a much better initial guess for the next iteration and covers a user specified range of admissible k.

The tremendous popularity of k-means algorithm has brought to life many other extensions and modifications. Mahalanobis distance can be used to cover hyperellipsoidal clusters [325]. The maximum of intracluster variances, instead of the sum, can serve as an objective function [200]. Generalizations that incorporate categorical attributes are also known: the term *prototype* is used in this context [234] instead of the term *centroid*. Modifications that construct clusters of balanced size are discussed in Sect. 7.1.

4 Density-Based Partitioning

An open set in the Euclidean space (actually in topological space) can be divided into its connected components. The implementation of this idea for partitioning a discrete set of points requires concepts of density, connectivity, and boundary. Definitions of these concepts are closely related to a point's

Fig. 3. Irregular shapes

nearest neighbors. A cluster, defined as a connected dense component, grows in any direction that density leads. Therefore, density-based algorithms are capable of discovering clusters of arbitrary shapes. Also this provides a natural protection against outliers. Figure 3 illustrates some cluster shapes that present a problem for partitioning relocation clustering (e.g., k-means), but are handled properly by density-based algorithms. Density-based algorithms are scalable. These outstanding properties come along with certain inconveniences. One inconvenience is that a single dense cluster consisting of two adjacent areas with significantly different densities (both higher than a threshold) is not very informative. Another drawback is a lack of interpretability. An excellent introduction to density-based methods is contained in [217].

Because density-based algorithms require a metric space, the natural setting for them is *spatial* data [218, 287]. To make computations feasible, some index of data is constructed (such as an R^* tree). Index construction is a topic of active research. Classic indices were effective only with reasonably low-dimensional data.

There are two major approaches for density-based methods. The first approach pins density to a training data point and is reviewed in Sect. 4.1. Representative algorithms include DBSCAN, GDBSCAN, OPTICS, and DB-CLASD. The second approach pins density to a point in the attribute space and is explained in Sect. 4.2. It is represented by the algorithm DENCLUE that is less affected by data dimensionality.

4.1 Density-Based Connectivity

Crucial concepts in this section are *density* and *connectivity*, both measured in terms of local distribution of nearest neighbors.

The algorithm DBSCAN (Density Based Spatial Clustering of Applications with Noise) by Ester et al. [152] targeting low-dimensional spatial data is the major representative in this category. Two input parameters ϵ and *MinPts* are used to introduce:

1. An ϵ-*neighborhood* $N_\epsilon(x) = \{y \in X \mid \text{dist}(x,y) \leq \epsilon\}$ of the point x,
2. A *core object*, a point with a $|N_\epsilon(x)| \geq MinPts$,
3. A notion of a point y *density-reachable* from a core object x (a sequence of core objects between x and y exists such that each belongs next to an ϵ-neighborhood of its predecessor),
4. A definition of *density-connectivity* between two points x, y (they should be density-reachable from a common core object).

Density-connectivity is an equivalence relation. All the points reachable from core objects can be factorized into maximal connected components serving as clusters. The points not connected to any core point are declared to be outliers and are not covered by any cluster. The noncore points inside a cluster represent its *boundary*. Finally, core objects are *internal* points.

DBSCAN processing is independent of data ordering. Obviously, effective computing of ϵ-neighborhoods presents a problem. For low-dimensional spatial data effective (meaning $O(log(N))$ rather than $O(N)$ fetches per search) indexing schemes exist. The algorithm DBSCAN relies on R^* tree indexing [293]. Therefore, in low-dimensional spatial data the theoretical complexity of DBSCAN is $O(Nlog(N))$. Experiments confirm slightly superlinear runtime.

Note that DBSCAN relies on ϵ-neighborhoods and on frequency counts within such neighborhoods to define core objects. Many spatial databases contain extended objects such as polygons instead of points. Any reflexive and symmetric predicate (for example, "two polygons have a nonempty intersection") suffices to define a "neighborhood." Additional measures (as intensity of a point) can be used instead of a simple count as well. These two generalizations lead to the algorithm GDBSCAN [383], which uses the same two parameters as DBSCAN.

With regard to the two parameters ϵ and $MinPts$, there is no straightforward way to fit them to data. Moreover, different parts of the data set could require different parameters – the problem discussed earlier in conjunction with CHAMELEON. The algorithm OPTICS (Ordering Points To Identify the Clustering Structure) developed by Ankerst et al. [23] adjusts DBSCAN to address this issue. OPTICS builds an augmented ordering of data, which is consistent with DBSCAN, but goes a step further: keeping the same two parameters ϵ and $MinPts$, OPTICS covers a spectrum of all different $\epsilon' \leq \epsilon$. The constructed ordering can be used automatically or interactively. With each point, OPTICS stores only two additional fields, the so-called core and reachability-distances. For example, the core distance is the distance to $MinPts'$ nearest neighbor when it does not exceed ϵ, or undefined otherwise. Experimentally, OPTICS exhibits runtime roughly equal to 1.6 of DBSCAN runtime.

While OPTICS can be considered an extension of DBSCAN in the direction of different local densities, a more mathematically sound approach is to consider a random variable equal to the distance from a point to its nearest neighbor and to learn its probability distribution. Instead of relying on user-defined parameters, a possible conjecture is that each cluster has its own typical distance-to-nearest-neighbor scale. The goal is to discover these scales. Such a *nonparametric* approach is implemented in the algorithm DB-CLASD (Distribution Based Clustering of Large Spatial Databases) [446]. Assuming that points inside each cluster are uniformly distributed (which may or may not be realistic), DBCLASD defines a cluster as a nonempty arbitrary shape subset in X that has the expected distribution of distance to the nearest neighbor with a required confidence and is the *maximal connected* set with this quality. DBCLASD handles spatial data, of the form used to describe a minefield, for example. The χ^2 test is used to check a distribution requirement, with the standard consequence that each cluster has to have at least 30 points. Regarding connectivity, DBCLASD relies on a *grid-based* approach to generate cluster-approximating polygons. The algorithm contains provisions for handling real databases with noise and implements *incremental* unsupervised learning. Two techniques are used. First, assignments are not final: points can change cluster membership. Second, certain points (noise) are not assigned, but are tried later. Therefore, once incrementally fetched, points can be revisited internally. DBCLASD is known to run faster than CLARANS by a factor of 60 on some examples. In comparison with the much more efficient DBSCAN, it can be 2–3 times slower. However, DBCLASD requires no user input, while an empirical search for appropriate parameters requires several DBSCAN runs. In addition, DBCLASD discovers clusters of different densities.

4.2 Density Functions

Hinneburg and Keim [229] shifted the emphasis from computing densities pinned to data points to computing density functions defined over the attribute space. They proposed the algorithm DENCLUE (DENsity-based CLUstEring). Along with DBCLASD, it has a firm mathematical foundation. DENCLUE uses a *density function*,

$$f^D(x) = \sum_{y \in D(x)} f(x, y),$$

which is the superposition of several *influence functions*. When the f-term depends on $x - y$, the formula can be recognized as a convolution with a kernel. Examples include a *square wave* function $f(x, y) = \theta\left(\|x - y\|/\sigma\right)$ equal to 1, if the distance between x and y is less than or equal to σ, and a Gaussian influence function $f(x, y) = \exp\left(-\|x - y\|^2/2\sigma\right)$. This provides a high level of generality: the first example leads to DBSCAN, and the second to k-means clusters! Both examples depend on the parameter σ. Restricting the summation to $D = \{y : \|x - y\| < k\sigma\} \subset X$ enables a practical implementation. DENCLUE concentrates on local maxima of density functions called *density attractors* and uses a gradient hill-climbing technique to find them. In addition to *center-defined* clusters, *arbitrary-shape* clusters are defined as unions of local shapes along sequences of neighbors whose local densities are no less than a prescribed threshold ξ. The algorithm is stable with respect to outliers. The authors show how to choose parameters σ and ξ. DENCLUE scales well, because at its initial stage it builds a *map* of hypercubes with edge length 2σ. For this reason, the algorithm can be classified as a grid-based method. Applications include high-dimensional multimedia and molecular biology data. While no clustering algorithm could have less than $O(N)$ complexity, the runtime of DENCLUE scales with N sublinearly! The explanation is that though all the points are fetched, the bulk of the analysis in the clustering stage involves only points in highly populated areas.

5 Grid-Based Methods

In Sect. 4, the crucial concepts of density, connectivity, and boundary were used that required elaborate definitions given purely in terms of distances between the points. Another way of dealing with these concepts is to inherit the topology from the underlying attribute space. To limit the amount of computations, multirectangular segments are considered (in like fashion to grids in analysis). Recall that a *segment* (also *cube, cell,* or *region*) is a direct Cartesian product of individual attribute subranges. Because some binning is usually adopted for numerical attributes, methods that partition the space are frequently called grid-based methods. The elementary segment, whose sides correspond to single-bins or single-value subranges, is called a *unit*.

In this section our attention is shifted from data to space partitioning. Data partitioning is induced by a point's membership in segments resulting from space partitioning, while space partitioning is based on grid characteristics accumulating from input data. One advantage of this indirect handling (data $\rightarrow$ grid data $\rightarrow$ space partitioning $\rightarrow$ data partitioning) is that accumulation of grid data makes grid-based clustering techniques independent of data ordering in contrast with relocation methods. Also notice that while density-based partitioning methods work best with numerical attributes, grid-based methods work with attributes of various types.

To some extent, the grid-based methodology reflects a technical point of view. The category is eclectic: it contains both partitioning and hierarchical algorithms. The algorithm DENCLUE from Sect. 4.2 uses grids at its initial stage and so it can be partially classified as grid based. The very important grid-based algorithm CLIQUE and its descendent, algorithm MAFIA, are presented in Sect. 9. In this section we review algorithms that rely on grid-based techniques as their principal instrument.

Schikuta and Erhart [389] introduced BANG clustering that summarizes data over the segments. The segments are stored in a special BANG structure

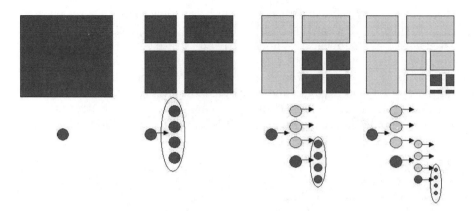

Fig. 4. Algorithm STING

that is a grid-directory incorporating different scales. Adjacent segments are considered neighbors. If a common face has maximum dimension they are called nearest neighbors. More generally, neighbors of degree (dimensions of a common face) between 0 and $d-1$ can be defined. The density of a segment is defined as a ratio between the number of points in that segment and its volume. From the grid directory, a hierarchical clustering (a dendrogram) is calculated directly. The algorithm BANG improves the similar earlier hierarchical algorithm GRIDCLUST [388].

The algorithm STING (STatistical INformation Grid-based method) developed by Wang et al. [437] works with numerical attributes (spatial data) and is designed to facilitate "region oriented" queries. STING assembles summary statistics in a hierarchical tree of nodes that are grid cells. Figure 4 illustrates the proliferation of cells in two-dimensional space and the construction of the corresponding tree. Each cell has four (default) children. Each cell stores a point count, and attribute-dependent measures: mean, standard deviation, minimum, maximum, and distribution type. Measures are accumulated starting from bottom-level cells. They are further aggregated to higher-level cells (e.g., minimum is equal to a minimum among the children minimums). Aggregation works fine for each measure except those of a distribution type, in which case the χ^2-test is used after bottom cell distribution types are handpicked. When the cell tree is constructed (in $O(N)$ time), certain cells are identified and connected in clusters similar to DBSCAN. If the cell tree has K leaves, the cluster construction phase depends on K and not on N. STING is parallelizable and allows for multiresolution, but defining an appropriate level of granularity is not straightforward. STING+ [439] is an enhancement of STING that targets dynamically evolving spatial databases using a hierarchical cell organization similar to its predecessor. In addition, STING+ enables *active* data mining by supporting user-defined *trigger conditions* (e.g., "at least 10 cellular phones are in use per square mile in a region of at least 10 square miles," or "usage drops by 20% in a described region"). The related measures ("subtriggers") are stored and updated over the cell tree. They are suspended until the trigger *fires* with user-defined action. Four types of conditions are supported: absolute and relative conditions on regions (a set of adjacent cells), and absolute and relative conditions on certain attributes.

The algorithm WaveCluster [394] also works with numerical attributes and supports an advanced multiresolution. It is known for the following outstanding properties:

- High-quality clusters,
- The ability to work well in relatively high-dimensional spatial data,
- Successful outlier handling,
- $O(N)$ complexity (it, however, exponentially grows with the dimension).

WaveCluster is based on ideas of signal processing, in that it applies wavelet transforms to filter the data. Note that high-frequency parts of a signal correspond to boundaries, while low-frequency high-amplitude parts of a signal correspond to clusters' interiors. The wavelet transform provides useful filters. For example, a hat-shaped filter forces dense areas to serve as attractors and simultaneously suppresses less dense boundary areas. After getting back from the signal to the attribute space this makes clusters sharper and eliminates outliers. WaveCluster proceeds in stages: (1) bin every dimension, assign points to units, and compute unit summaries, (2) apply discrete wavelet transform to the accumulated units, (3) find connected components (clusters) in the transformed attribute space (corresponding to a certain level of resolution), and (4) assign points.

The hierarchy of grids allows definition of the *Hausdorff Fractal Dimension* (HFD) [387]. The HFD of a set is the negative slope of a log–log plot of the number of cells $Cell(r)$ (occupied by a set) as a function of a grid size r. A fast (*box counting*) algorithm to compute HFD was introduced in [314]. The concept of HFD is fundamental to the FC (fractal clustering) algorithm by Barbara and Chen [49] dealing with numeric attributes. FC works with several layers of grids. The cardinality of each dimension increases four times with each next layer. Although only occupied cells are kept to save memory, memory usage is still a significant problem. FC starts by initializing k clusters. The initialization threshold and a data sample are used at this stage to come up with an appropriate k. Then FC scans the full data incrementally attempting to add the incoming point to a cluster so as to minimally increase its HFD. If the smallest increase exceeds the threshold τ, a point is declared an outlier. The FC algorithm has some appealing properties:

- Incremental structure (batches of data are fetched into main memory),
- Suspendable nature always ready for online assignments,
- Ability to discover clusters of irregular shapes,
- $O(N)$ complexity.

as well as a few drawbacks:

- Data order dependency,
- Strong dependency on cluster initialization,
- Dependency on parameters (threshold used in initialization, and τ).

6 Co-occurrence of Categorical Data

This section focuses on clustering categorical data. Such data are often related to transactions involving a finite set of elements, or items, in a common item universe. For example, market basket data have this form. Every transaction is a set of items that can be represented in a point-by-attribute format, by enumerating all items j, and by associating with a transaction binary attributes that indicate whether the j-item belongs to a transaction or not. Such a representation is sparse, and high dimensional: two random transactions will in general have very few items in common. This is why the similarity (Sect. 10.4) between any two transactions is usually measured with the Jaccard coefficient $\text{sim}(T_1, T_2) = |T_1 \bigcap T_2| / |T_1 \bigcup T_2|$. In this situation, conventional clustering methods based on cosine similarity measures do not work well. However, because categorical/transactional data are important in customer profiling, assortment planning, Web analysis, and many other applications, different clustering methods founded on the idea of *co-occurrence* have been developed.

The algorithm ROCK (Robust Clustering algorithm for Categorical Data) [208] deals with categorical data and has many features in common with the hierarchical clustering algorithm CURE discussed in Sect. 2. That is, ROCK:

- Does hierarchical agglomerative clustering
- Continues agglomeration until a specified number k of clusters are constructed
- Uses data sampling in the same way as CURE does

ROCK defines a neighbor of a point x as a point y such that $\text{sim}(x, y) \geq \theta$ for some threshold θ, and it defines $\text{link}(x, y)$ between two points x, y as the number of neighbors they have in common. Clusters consist of points with a high degree of connectivity: $\text{link}(x, y)$ for pairs of points inside a cluster is on average high. ROCK utilizes the objective function

$$E = \sum_{j=1:k} |C_j| \times \sum_{x,y \in C_j} link(x, y) / |C_j|^{1+2f(\theta)},$$

where $f(\theta)$ is a data-dependent function. E represents a specifically normalized intraconnectivity measure.

To put this formula into perspective, we note that linkage metrics normalize the aggregate measures (combining measures associated with each edge) by the number of edges. For example, the average link metric is the sum of distances between each point in C_i and each point in C_j divided by the (normalization) factor $L = |C_i| \cdot |C_j|$. The value L can be rationalized on a more general level. If the expected number of edges per cluster is $|C|^\beta$, $\beta \in [1, 2]$, then the aggregate intercluster similarity has to be normalized by the factor $(|C_i| + |C_j|)^\beta - |C_i|^\beta - |C_j|^\beta$ representing the number of intercluster edges. The average link normalization factor L corresponds to $\beta = 2$, i.e., complete connectivity. The ROCK objective function uses the same idea but fits it with parameters. Whether an obtained model fits particular data is an open question. To facilitate fitting the data, ROCK relies on an input parameter θ and on a function $f(\theta)$. Frequently, different regions of data have different properties, and a global fit is therefore impossible. ROCK has a complexity of

$$O\left(c_m N_{\text{sample}} + N_{\text{sample}}^2 \log(N_{\text{sample}})\right),$$ where the coefficient c_m is a product of average and maximum number of neighbors.

The algorithm SNN (Shared Nearest Neighbors) [150] blends a density-based approach with the "shared nearest neighbors" idea of ROCK. SNN sparsifies the similarity matrix (therefore, unfortunately resulting in $O\left(N^2\right)$ complexity) by only keeping K-nearest neighbors. The idea of using shared nearest neighbors in clustering was suggested by Jarvis [247], see also [204], long ago.

The algorithm CACTUS (Clustering Categorical Data Using Summaries) by Ganti et al. [181] looks for hyperrectangular clusters (called *interval regions*) in point-by-attribute data with categorical attributes. In our terminology such clusters are segments. CACTUS is based on the idea of co-occurrence for attribute-value pairs. A uniform distribution within the range of values for each attribute is assumed. Two values a, b of two different attributes are *strongly connected* if the number of data points having both a and b is larger than the frequency expected under an independence assumption by a user-defined margin α. This definition is extended to subsets A, B of two different attributes (each value pair $a \in A, b \in B$ has to be strongly connected), to segments (each two-dimensional projection is strongly connected), and to the similarity of pair of values of a single attribute via connectivity to other attributes. The cluster is defined as the maximal strongly connected segment having at least α times more elements than expected from the segment under the attribute independence assumption. CACTUS uses data summaries to generate all the strongly connected value pairs. As a second step, a heuristic is used to generate maximum segments. The complexity of the summarization phase is $O\left(cN\right)$, where the constant c depends on whether all the attribute-value summaries fit in memory (one data scan) or not (multiple data scans).

Clustering transactional data becomes more difficult when the size of item universe grows. Here we have a classic case of low separation in high-dimensional space (Sect. 9). With categorical data, the idea of auxiliary items or more generally of grouping categorical values gained popularity. This is very similar to the idea of coclustering (Sect. 9.3). This attribute value clustering, done as a preprocessing step, becomes the major concern, while the subsequent data clustering becomes a lesser issue.

The work of Han et al. [215] exemplifies this approach for transactional data. After items are clustered (a major step), a very simple method to cluster transactions themselves is used: each transaction T is assigned to a cluster C_j of items having most in common with T, as defined by a function $|T \cap C_j| / |C_j|$. Other choices are possible, but again the primary objective is to find item groups. To achieve this, *association rules* and *hypergraph* machineries are used. First, frequent item sets are generated from the transactional data. A hypergraph $H = (V, E)$ is associated with the item universe. Vertices V are items. In a common graph, pairs of vertices are connected by edges. In a hypergraph several vertices are connected by hyperedges. Hyperedge $e \in E$ in H corresponds to a frequent item set $\{\nu_1, \ldots, \nu_s\}$ and has a *weight* equal to an average of confidences among all association rules involving this item set. We thereby transform the original problem into hyper-graph partitioning problem. A solution to the problem of k-way partitioning of a hypergraph H is provided by algorithm HMETIS [258].

Gibson et al. [188] introduced the algorithm STIRR (Sieving Through Iterated Reinforcement), which deals with co-occurrence phenomenon for d-dimensional categorical objects, *tuples*. STIRR uses a beautiful technique from functional analysis. Define *configurations* as weights $w = \{w_\nu\}$ over all different values ν for each of d categorical attributes. We wish to define transformations over these weights. To do so, assume that a value ν belongs to the first attribute, and consider all data tuples of the form $x = (\nu, u_2, \ldots, u_d)$. Then we define a weight update $w'_\nu = \sum_{x, \nu \in x} z_x$, where $z_x = \Phi(w_{u_2}, \ldots, w_{u_d})$. This weight update depends on a *combining operator* Φ. An example of a combining operator is $\Phi(w_2, \ldots, w_d) = w_2 + \cdots + w_d$. To get a transformation, an update is followed by a renormalization of weights among the values of each attribute. This purely technical transformation reflects a fundamental idea of this section: weights of the items propagate to other items with which the original items co-occur. If we start with some weights and propagate them several times, then assuming that propagation process stabilizes, we get some balanced weights. The major iteration scans the data X and results in one transformation. Function f can be considered as a *dynamic system* $w_{\text{new}} = f(w)$ (nonlinear, if Φ is nonlinear).

STIRR relies deeply ideas from *spectral graph partitioning*. For a linear dynamic system defined over the graph, a reorthogonalization Gram–Schmidt process can be engaged to compute its eigenvectors that introduce negative weights. The few first nonprincipal eigenvectors (nonprincipal basins) define a graph partitioning corresponding to positive/negative weights. The process works as follows: a few weights (configurations) $w^q = \{w^q_\nu\}$ are initialized. A major iteration updates them, $w^q_{\text{new}} = f(w^q)$, and new weights are reorthogonalized. The process continues until a fixed point of the dynamic system is achieved. Nonprincipal basins are analyzed. In STIRR a dynamic system instead of association rules formalizes co-occurrence. Additional references related to spectral graph partitioning can be found in [188]. As the process does not necessarily *converge*, further progress is related to the modification of the dynamic system that guarantees the convergence [461].

7 Other Clustering Techniques

A number of other clustering algorithms have been developed. *Constraint-based clustering* deals with the use of specific application requirements. *Graph partitioning* is an independent active research field. Some algorithms have theoretical significance or are mostly used in applications other than data mining.

7.1 Constraint-Based Clustering

In real-world applications customers are rarely interested in unconstrained solutions. Clusters are frequently subjected to some problem-specific limitations that make them suitable for particular business actions. Finding clusters satisfying certain limitations is the subject of active research; for example, see a survey by Han et al. [218]. The framework for constraint-based clustering is introduced in [425]. Their taxonomy of clustering constraints includes constraints on individual objects (e.g., customer who recently purchased) and parameter constraints (e.g., number of clusters) that can be addressed through preprocessing or external cluster parameters. Their taxonomy also includes constraints on individual clusters that can be described in terms of bounds on aggregate functions (min, avg, etc.) over each cluster. These constraints are essential, because they require a new methodology. In particular, an *existential constraint* is a lower bound on a count of objects of a certain subset (i.e., frequent customers) in each cluster. Iterative optimization used in partitioning clustering relies on moving objects to their nearest cluster representatives. This movement may violate such constraints. The authors developed a methodology of how to incorporate constraints in the partitioning process.

The most frequent requirement is to have a minimum number of points in each cluster. Unfortunately, the popular k-means algorithm sometimes provides a number of very small (in certain implementations empty) clusters. Modification of the k-means objective function and the k-means updates to incorporate lower limits on cluster volumes is suggested by Bradley et al. [86]. This modification includes soft assignments of data points with coefficients subject to linear program requirements. Banerjee and Ghosh [44] presented another modification to the k-means algorithm. Their objective function corresponds to an isotropic Gaussian mixture with widths inversely proportional to the numbers of points in the clusters. The result is the *frequency sensitive* k-means. Still another approach to building balanced clusters is to convert the task into a graph-partitioning problem [409].

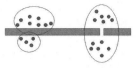

No obstacles   River with the bridge

Fig. 5. COD

An important constraint-based clustering application is clustering two-dimensional spatial data in the presence of obstacles. Instead of regular Euclidean distance, a length of the shortest path between two points can be used as an obstacle distance. The COD (Clustering with Obstructed Distance) algorithm [426] deals with this problem. COD is illustrated in Fig. 5, where we show the difference in constructing three clusters in the absence of any obstacle (left) and in the presence of a river with a bridge (right).

7.2 Graph Partitioning

We now briefly discuss the active research field of graph partitioning. Graphs frequently exhibit a clustering tendency and are important in many applications (e.g., VLSI design). The domain of graph clustering has a methodology of its own. A graph can be partitioned by simply deleting (cutting) some of its edges. Minimal number of cuts is desirable, but this is known to give very unbalanced clusters. Therefore a min-cut objective function is usually modified. Different modifications of this objective function are evaluated in [462].

Exact optimization of any min-cut modification is NP hard. Much progress was achieved with the introduction of an algebraic technique related to a second eigenvector of a Laplacian operator, which is simply a graph adjacency matrix with a changed main diagonal [161]. Since then, spectral methods have been under constant development [53, 139, 206, 212, 348]. A general discussion on the topic can be found in [431]. For spectral methods in the context of document clustering, see [128].

Another approach to graph partitioning is based on the idea of graph flows. A survey of this research is presented in [309]. A specific Markov cluster algorithm based on the simulation of (stochastic) flow is described in [428].

7.3 Relation to Supervised Learning

Both Forgy's k-means implementation and EM algorithms are iterative optimizations. Both initialize k models and then engage in a series of two-step iterations that: (1) reassign (*hard* or *soft*) data points, and then (2) update a combined model. This process can be generalized to a framework relating clustering with predictive mining [253]. A model update can be considered as training a predictive classifier based on current assignments serving as the target attribute values supervising the learning. Points are reassigned according to the forecast of the recently trained classifier.

Liu et al. [320] suggested another elegant connection to supervised learning. They considered a binary target attribute defined as *Yes* on points subject to clustering, and defined as *No* on synthetically generated points uniformly distributed in the attribute space. A decision tree classifier is trained on the full data. *Yes*-labeled leaves correspond to clusters of input data. The algorithm CLTree (CLustering based on decision Trees) resolves the challenges of populating the input data with artificial *No* points by: (1) adding points gradually following the tree construction; (2) making this process virtual (without physical additions to input data); and (3) simulating the uniform distribution in higher dimensions.

7.4 Gradient Descent and Artificial Neural Networks

Soft reassignments make a lot of sense if the k-means objective function is modified to incorporate "fuzzy errors" (similar to EM), i.e., to account for distances not only to the closest centroids, but also to more distant centroids:

$$E'(C) = \sum_{i=1:N} \sum_{j=1:k} \|x_i - c_j\|^2 \, \omega_{ij}^2.$$

The probabilities ω are defined based on Gaussian models. This makes the objective function differentiable with respect to means and allows application of general *gradient descent* methods. Marroquin and Girosi [327] presented a detailed introduction to this subject in the context of *vector quantization*. The gradient decent method in k-means is known as LKMA (Local k-Means Algorithm). At each iteration t, LKMA updates means c_j^t,

$$c_j^{t+1} = c_j^t + a_t \sum_{i=1:N} (x_i - c_j^t)w_{ij}^2, \quad \text{or} \quad c_j^{t+1} = c_j^t + a_t(x_i - c_j^t)w_{ij}^2,$$

in the direction of gradient descent. In the second case index i is selected randomly between 1 and N. The scalars a_t satisfy certain monotone asymptotic behavior and converge to 0, and the coefficients w are defined through ω [81]. Such updates are also used in the context of *artificial neural network* (ANN) clustering, specifically SOM (Self-Organizing Map) [283]. SOM is popular in vector quantization and is described in the monograph [284]. We do not elaborate on SOM here except for two remarks: (1) SOM uses an incremental approach – points (patterns) are processed one at a time; and (2) SOM allows one to map centroids into the two-dimensional plane making it one of a very few successful clustering visualization techniques. In addition to SOM, other ANN developments, such as *adaptive resonance theory* [97], are related to clustering. For further discussion, see [244].

7.5 Evolutionary Methods

Substantial information on the application of *simulated annealing* in the context of partitioning (main focus) or hierarchical clustering has been accumulated, including the algorithm SINICC (SImulation of Near-optima for Internal Clustering Criteria) [91]. The perturbation operator used in general annealing has a simple meaning in clustering, namely the relocation of a point from its current to a new randomly chosen cluster. SINICC also tries to address the interesting problem of choosing the most appropriate objective function. It has a real application – surveillance monitoring of ground-based entities by airborne and ground-based sensors. Similar to simulating annealing is the so-called *tabu* search [16].

Genetic Algorithms (GA) [197] are also used in cluster analysis. An example is the GGA (Genetically Guided Algorithm) for fuzzy and hard k-means [213]. Sarafis et al. [384] applied GA in the context of k-means objective functions. A population is a set of "k-means" systems represented by grid segments instead of centroids. Every segment is described by d rules (genes), one per attribute range. The population is improved through mutation and crossover specifically devised for these rules. Unlike normal k-means, GGA clusters can have different sizes and elongations; however, shapes are restricted to just k segments. GA have also been applied to clustering of categorical data using generalized entropy to define dissimilarity [117].

Evolutionary techniques rely on parameters to empirically fit data and have high computational costs that limit their application in data mining. However, combining evolutionary techniques with other strategies (e.g., the generation of initial partitions for k-means) has been attempted [36,37]. Use of GA with variable length genomes to simultaneously improve k-means centroids and k itself [305] also compares favorably with running multiple k-means to determine k, because changes in k happen before full convergence is achieved.

7.6 Other Developments

Some clustering methods do not fit in our classification. For example, for two-dimensional spatial data (such as is found in GIS databases) the algorithm AMOEBA [154] uses Delaunay diagrams (the dual of Voronoi diagrams) to represent data proximity and has $O(N \log(N))$ complexity.

Harel and Koren [221] suggested an approach related to agglomerative hierarchical graph clustering that successfully finds local clusters in two-dimensional. Consider a connectivity graph $G(X, E)$. The graph can be made sparse through the use of Delaunay diagrams or by keeping with any point only its K-nearest neighbors. The method relies on a *random walk* to find separating edges F so that clusters become connected components of $G(X, E - F)$.

8 Scalability and VLDB Extensions

Clustering algorithms face scalability problems in terms of both computing time and memory requirements. In data mining, reasonable runtime and ability to run in limited amounts of main memory become especially important. There have been many interesting attempts to extend clustering to very large databases (VLDB), which can be divided into:

- Incremental mining,
- Data squashing,
- Reliable sampling.

The algorithm DIGNET [419, 440] (compare with "the leader" clustering algorithm in [224]) is an example of incremental unsupervised learning. That is, DIGNET handles one data point at a time, and then discards it. DIGNET uses k-means cluster representation without iterative optimization. Centroids are instead *pushed* or *pulled* depending on whether they lose or win each next coming point. Such online clustering needs only one pass over the data, but strongly depends on data ordering, and can result in poor-quality clusters. However, it handles outliers, clusters can be dynamically created or discarded, and the training process is resumable. This makes DIGNET appealing for dynamic VLDB. The clusters resulting from DIGNET may serve as initial guesses for other algorithms.

Data *squashing* techniques scan data to compute certain data summaries (*sufficient statistics*) [146]. The summaries are then used instead of the original data for clustering. Here the most important role belongs to the algorithm BIRCH (Balanced Iterative Reduction and Clustering using Hierarchies) developed by Zhang et al. [459, 460]. BIRCH has had a significant impact on the overall direction of scalability research in clustering. BIRCH creates a height-balanced tree of nodes that summarizes data by accumulating its zero, first, and second moments. A node, called *Cluster Feature* (CF), is a small, tight cluster of numerical data. The aforementioned tree resides in main memory. A new data point descends along the tree to the closest CF leaf. If it fits the leaf well, and if the leaf is not overcrowded, CF statistics are incremented for all nodes from the leaf to the root. Otherwise a new CF is constructed. Because the maximum number of children per node (*branching factor*) is limited, one or several splits can happen. When the tree reaches the assigned memory size, it is rebuilt and a threshold controlling whether a new point is assigned to an existing leaf or starts a new leaf is updated to a coarser one. The outliers are saved to a file and refitted gradually during the tree rebuilds. The final leaves constitute input to any algorithm of choice. The fact that a CF tree is balanced allows for log-efficient search.

BIRCH depends on parameters that control CF tree construction (branching factor, maximum of points per leaf, leaf threshold) and also on data ordering. When the tree is constructed (one data pass), it can be additionally condensed in the optional second phase to further fit the desired input cardinality of a postprocessing clustering algorithm. Next, in the third phase a global clustering of CF (considered as individual points) happens. Finally, certain irregularities (for example, identical points being assigned to different CFs) can be resolved in an optional fourth phase. This fourth phase makes one or more passes through data reassigning points to the best possible clusters, as k-means does. The overall complexity of BIRCH is $O(N)$. BIRCH has been extended to handle mixed numerical and categorical attributes [111].

According to Bradley et al. [85], a full interface between VLDB and relocation clustering (e.g., k-means) has the following requirements. The algorithm must:

- Make one full scan of the data, or less in case of early termination,
- Provide online intermediate solutions,
- Be suspendable, stoppable, and resumable,
- Be able to incorporate additional data incrementally,
- Be able to work in a prescribed amount of memory,
- Utilize different scanning modes (sequential, index, sample),
- Be able to operate in forward-only cursor over a view of database.

The authors suggest data compression that accumulates sufficient statistics like BIRCH does, but makes it in phases. Points that are compressed over the primary stage are discarded. They can be attributed to their clusters with very high confidence even if other points would shift. The rest is taken care of in the secondary phase, which tries to find dense subsets by the k-means method with higher than requested k. Violators of this stage are still kept in the retained set (RT) of singletons to be analyzed later.

BIRCH-like preprocessing relies on vector-space operations. In many applications, objects (for example, strings) belong to a metric space. In other words, all we can do with data points is compute distances between them. Ganti et al. [182] proposed a BIRCH-type data squashing technique called BUBBLE that works with VLDB in metric spaces. Each leaf of the BUBBLE-tree is characterized by: (1) its number of points, (2) its medoid (or *clustroid*), i.e., that point with the least squared distance between it and all other points belonging to the leaf, and (3) its radius, which is the average distance between the medoid and the other leaf points.

The problem is how to insert new points in the absence of a vector structure. BUBBLE uses a heuristic that relates to a distance preserving embedding of leaf points into a low-dimensional Euclidean vector space. Such embedding is known as an isometric map in geometry and as a multidimensional scaling in statistics. An analogy can also be made with embeddings used in support vector machines. While Euclidean distance (used in BIRCH) is cheap, the computation of a distance in an arbitrary metric space (for example, edit distance for strings) can be expensive. Meanwhile, every insertion requires calculating distances to all the nodes descending to a leaf. The sequel algorithm BUBBLE-FM handles this difficulty. BUBBLE-FM reduces the computation by using *approximate* isometric embedding, using the algorithm FastMap [158].

In the context of hierarchical density-based clustering in VLDB, Breunig et al. [89] analyzed data reduction techniques such as sampling and BIRCH summarization and noticed that they result in deterioration of cluster quality. To cure this, they approached data reduction through accumulation of *data bubbles* that are summaries of local information about distances and nearest neighbors. A data bubble contains an *extent* (the distance from a bubble's representative to most points in X), and the array of distances to each of its *MinPts* nearest neighbors. Data bubbles are then used in conjunction with the algorithm OPTICS (see Sect. 4.1).

Grid methods also generate data summaries, though their summarization phase relates to units and segments and not to CFs. Therefore, they are scalable.

Many algorithms use old-fashioned sampling with or without rigorous statistical reasoning. Sampling is especially handy for different initializations as in CLARANS (Sect. 3.2), fractal clustering (Sect. 5), or k-means [84]. Note that when clusters are constructed using any sample, assigning the whole data set to the most appropriate clusters adds at least the term $O(N)$ to the overall complexity.

Sampling has received new life with the adoption by the data mining community of a special uniform test to control its adequacy. This test is based on Hoeffding or Chernoff bounds [344] and states that, independent of the distribution of a real-valued random variable Y, $0 \leq Y \leq R$, the average of n independent observations lies within ϵ of the actual mean,

$$\left| \bar{Y} - \frac{1}{n} \sum_{j=1:n} Y_j \right| \leq \epsilon$$

with probability $1 - \delta$ as soon as,

$$\epsilon = \sqrt{R^2 \ln(1/\delta)/2n}.$$

These bounds are used in the clustering algorithm CURE [207] and in the development of scalable decision trees in predictive mining [236]. In the context of balanced clustering, a statistical estimation of sample size is provided in [44]. Due to their nonparametric nature, the bounds are useful in a variety of applications.

9 Clustering High-Dimensional Data

The objects in data mining can have hundreds of attributes. Clustering in such high-dimensional spaces presents tremendous difficulty, much more so than in predictive learning. For example, a decision tree simply skips an irrelevant attribute in node splitting. Such attributes do not significantly affect Naïve Bayes either. In clustering, however, high dimensionality presents a dual problem. First, under whatever definition of similarity, the presence of irrelevant attributes eliminates any hope on *clustering tendency*. After all, searching for clusters where there are no clusters is a hopeless enterprise. While this could also happen with low-dimensional data, the likelihood of the presence and number of irrelevant attributes grows with dimension.

The second problem is the *dimensionality curse*, which is a loose way of speaking about a lack of data separation in a high-dimensional space. Mathematically, the nearest neighbor query becomes unstable, in that the distance to the nearest neighbor becomes indistinguishable from the distance to the majority of points [68]. This effect starts to be severe for dimensions greater than 15. Therefore, construction of clusters founded on the concept of proximity is doubtful in such situations. For interesting insights into complications of high-dimensional data, see [10].

In Sect. 9.1 we talk briefly about traditional methods of dimensionality reduction. In Sect. 9.2 we review algorithms that try to circumvent high dimensionality by building clusters in appropriate subspaces of the original attribute space. Such an approach makes perfect sense in applications, because it is better if we can describe data with fewer attributes. Still another approach that divides attributes into similar groups and comes up with good new derived attributes representing each group is discussed in Sect. 9.3.

9.1 Dimensionality Reduction

When talking about high dimensionality, how high is high? Many clustering algorithms depend on indices in spatial datasets to facilitate quick search of the nearest neighbors. Therefore, indices can serve as good proxies with respect to the performance impact of the *dimensionality curse*. Indices used in clustering algorithms are known to work effectively for dimensions below 16. For a dimension $d > 20$ their performance degrades gradually to the level of sequential search (though newer indices achieve significantly higher limits). Therefore, we can arguably claim that data with more than 16 attributes are high dimensional.

How large is the gap between a nonhigh dimension and a dimension in real-life applications? If we are dealing with a retail application, for example, 52-week sales volumes (52 attributes) represent a typical set of features, which is a special instance of the more general class of time series data. In customer profiling, dozens of item categories plus basic demographics result in at the least 50–100 attributes. Web clustering based on site contents results in 1,000–5,000 attributes (pages/contents) for modest Web sites. Biology and genomic data easily surpass 10,000 attributes. Finally, text mining and information retrieval routinely deal with many thousands of attributes (words or index terms). So, the gap is significant. Two general purpose techniques are used to combat high dimensionality: (1) *attribute transformation* and (2) *domain decomposition*.

Attribute transformations are simple functions of existing attributes. One example is a sum or an average roll-up for sales profiles or any OLAP-type data (e.g., monthly volumes). Because of the fine seasonality of sales such brute force approaches rarely work. In multivariate statistics, principal component analysis (PCA) is popular [251,326], but this approach is problematic due to poor interpretability. The singular value decomposition (SVD) technique is used to reduce dimensionality in information retrieval [60, 61] and statistics [180]. Low-frequency Fourier harmonics in conjunction with Parseval's theorem are successfully used in analysis of time series [14], as well as wavelets and other transformations [267].

Domain decomposition divides the data into subsets, *canopies* [329], using some inexpensive similarity measure, so that the high-dimensional computation happens over smaller datasets. The dimension stays the same, but the cost of computation is reduced. This approach targets the situation of high dimension, large data sets, and many clusters.

9.2 Subspace Clustering

Some algorithms adapt to high dimensions more easily than others. For example, the algorithm CACTUS (Sect. 6) adjusts well because it defines a cluster only in terms of a cluster's two-dimensional projections. In this section we cover techniques that are specifically designed to work with high dimensional data.

The algorithm CLIQUE (Clustering In QUEst) invented by Agrawal et al. [15] is fundamental for high-dimensional numerical data. CLIQUE combines the ideas of:

- Density and grid-based clustering,
- Induction through dimension similar to the *Apriori* algorithm,
- MDL principle to select appropriate subspaces,
- Interpretability of clusters in terms of DNF representation.

CLIQUE starts with the definition of a unit, an elementary rectangular cell in a subspace. Only units whose densities exceed a threshold τ are retained. A bottom-up approach of finding such units is applied. First, 1s units are found by dividing intervals into a grid of ξ equal-width bins. Both parameters τ and ξ are inputs to the algorithm. The recursive step from $q - 1$-dimensional units to q-dimensional units involves a self-join of the $q - 1$ units sharing *first* common $q - 2$ dimensions (Apriori reasoning in association rules). All the subspaces are sorted by their coverage and lesser-covered subspaces are pruned. A cut point between retained and pruned subspaces is selected based on the MDL principle. A cluster is defined as a maximal set of connected dense units and is represented by a DNF expression that specifies a finite set of maximal segments (called *regions*) whose union is the cluster. Effectively, CLIQUE performs attribute selection (it selects several subspaces) and produces a view of data from different perspectives. The result is a series of cluster systems in different subspaces. Such systems overlap. Thus, CLIQUE produces a description of the data rather than a partitioning. If q is the highest subspace dimension selected, the complexity of dense unit generations is $O(\text{const}^q + qN)$. Finding clusters is a task quadratic in terms of units.

The algorithm ENCLUS (ENtropy-based CLUStering) [108] follows the footsteps of CLIQUE, but uses a different criterion for a subspace selection. The criterion is derived from entropy-related considerations: the subspace spanned by attributes $A_1, \ldots, A_q$ with entropy $H(A_1, \ldots, A_q) < \omega$ (a threshold) is considered good for clustering. Indeed, a low-entropy subspace corresponds to a skewed distribution of unit densities. Any subspace of a good subspace is also good, since

$$H(A_1, \ldots, A_{q-1}) = H(A_1, \ldots, A_q) - H(A_q | A_1, \ldots, A_{q-1}) < \omega.$$

The computational costs of ENCLUS are high.

The algorithm MAFIA (Merging of Adaptive Finite Intervals) [196, 347] significantly modifies CLIQUE. MAFIA starts with one data pass to construct *adaptive grids* in each dimension. Many (1,000) bins are used to compute histograms by reading blocks of data into memory. The bins are then merged to come up with a smaller number of adaptive variable-size bins than CLIQUE. The algorithm uses a parameter α, called the cluster dominance factor, to select bins that are α-times more densely populated than average. These variable-size bins are $q = 1$ candidate dense units (CDUs). Then MAFIA proceeds recursively to higher dimensions (every time a data scan is involved). Unlike CLIQUE, when constructing a new q-CDU, MAFIA tries two $q - 1$-CDUs as soon as they share any (not only the first dimensions) $q-2$-face. This creates an order of magnitude more candidates. Adjacent CDUs are merged into clusters. Clusters that are proper subsets of other clusters are eliminated. Fitting the parameter α presents no problem (in practice, the default value 1.5 works fine) in comparison with the global density threshold used in CLIQUE. Reporting for a range of α in a single run is supported. If q is the highest dimensionality of CDU, the algorithm's complexity is $O(\text{const}^q + qN)$.

The algorithm OPTIGRID [230] partitions data based on divisive recursion by multidimensional grids. The authors present a very good introduction to the effects of high-dimension geometry. Familiar concepts, such as uniform distribution, become blurred for large d. OPTIGRID uses density estimations in the same way as the algorithm DENCLUE [229]. OPTIGRID focuses on separation of clusters by hyperplanes that are not necessarily axes parallel. To find such planes consider a set of *contracting* linear projectors (functionals) $P_1, \ldots, P_k, \|P_j\| \leq 1$ of the attribute space A at a one-dimensional line. For a density kernel of the form $K(x-y)$ utilized in DENCLUE and for a contracting projection, the density induced after projection is more concentrated. Each cutting plane is chosen so that it goes through the point of minimal density and discriminates two significantly dense half-spaces. Several cutting planes are chosen, and recursion continues with each subset of data.

The algorithm PROCLUS (PROjected CLUstering) [12] explores pairs consisting of a data subset $C \subset X$ and a subspace in an attribute space A. A subset–subspace pair is called a projected cluster, if a projection of C onto the corresponding subspace is a tight cluster. The number k of clusters and the average subspace dimension l are user inputs. The iterative phase of the algorithm deals with finding k good medoids, each associated with its subspace. A sample of data is used in a greedy hill-climbing technique. Manhattan distance divided by the subspace dimension is suggested as a useful normalized metric for searching among different dimensions. After the iterative stage is completed, an additional data pass is performed to refine clusters.

The algorithm ORCLUS (ORiented projected CLUSter generation) [13] uses a similar approach of projected clustering, but employs nonaxes parallel subspaces in high-dimensional space. In fact, both developments address a more generic issue: even in a low-dimensional space, different portions of data could exhibit clustering tendency in different subspaces (consider several nonparallel nonintersecting cylinders in three-dimensional space). If this is the case, any attribute selection is doomed. ORCLUS has a k-means like transparent model that defines a cluster as a set of points (i.e., a partition) that has low sum-of-squares of errors (*energy*) in a certain subspace. More specifically, for $x \in C$, and directions $E = \{e_1, \ldots, e_l\}$ (specific to C), the projection is defined as $\{x^T \cdot e_1, \ldots, x^T \cdot e_l\}$. The projection only decreases energy. SVD diagonalization can be used to find the directions (eigenvectors) corresponding to the lowest l eigenvalues of the covariance matrix. In reality, the algorithm results in X partitioning (the outliers excluded) into k clusters C_j together with their subspace directions E_j. The algorithm builds more than k clusters, with larger than l-dimensional E gradually fitting the optimal subspace and requested k. A suggestion for picking a good parameter l is based on experience with synthetic data.

Leaving any other comparisons aside, projected clusters provide data partitioning, while cluster systems constructed by CLIQUE overlap.

9.3 Coclustering

In OLAP attribute roll-ups can be viewed as representatives of the attribute groups. The interesting general idea of producing attribute groups in conjunction with clustering of points leads to the concept of *coclustering*. Coclustering is the simultaneous clustering of both points and their attributes. This approach partially reverses the struggle: to improve clustering of points based on their attributes, it tries to cluster attributes based on the points. So far we were concerned with grouping only rows of a matrix X. Now we intend to group its columns as well. This utilizes a canonical *duality* contained in the point-by-attribute data representation.

The idea of coclustering of data points and attributes is old [21, 224] and is known under the names *simultaneous clustering*, *bi-dimensional clustering*, *block clustering, conjugate clustering, distributional clustering*, and *information bottleneck method*. The use of duality for analysis of categorical data (dual or multidimensional scaling) also has a long history in statistics [351]. A similar idea of grouping items is also presented in Sect. 6. In this section we turn to numerical attributes. Assume that the matrix X has non-negative elements. Such matrices appear as *incidence, relational, frequency*, or *contingency* matrices. In applications it can reflect the intensity of a gene response in a tissue sample, the frequency of Web page visitation activity, or sales volume per store per category.

Govaert [203] researched simultaneous block clustering of the rows and columns of contingency tables. He also reviewed an earlier work on the subject. An advanced algebraic approach to coclustering based on bipartite graphs and their minimal cuts in conjunction with text mining was developed in [128]. This paper contains an excellent introduction to the relationships between simultaneous clustering, graph partitioning and SVD. A simple algorithm Ping-Pong [354] was suggested to find populated areas in a sparse binary matrix. Ping-Pong redistributes influence of columns on rows and vice versa by transversal connection through matrix elements (compare with algorithm STIRR mentioned earlier).

A series of publications deal with distributional clustering of attributes based on the informational measures of attribute similarity. Two attributes (two columns in matrix X) with exactly the same probability distributions are identical for the purpose of data mining, and so, one can be deleted. Attributes that have probability distributions that are close in terms of their Kullback–Leibler (KL) distance [298] can still be grouped together without much impact on information contained in the data. In addition, a natural derived attribute, the mixed distribution (a normalized sum of two columns), is now available to represent the group. This process can be generalized. The grouping simplifies the original matrix X into the compressed form $\bar{X}$. Attribute clustering is productive when it minimizes information reduction $R = I(X) - I(\bar{X})$, where $I(X)$ is mutual information contained in X [115]. Such attribute grouping is intimately related to Naïve Bayes classification in predictive mining [41].

The attribute-grouping technique just described is quite relevant to grouping words in text mining. In this context the technique was explored under the name *Information Bottleneck method* [421], and was used to facilitate agglomerative co-clustering of words in document clustering [400] and classification [401].

Berkhin and Becher [58] showed the deep algebraic connection between distributional clustering and k-means. They used k-means adaptation to KL-distance as a major iterative step in the algorithm SimplifyRelation that gradually coclusters points and attributes. This development has industrial application in Web analysis. Figure 6 shows how an original incidence matrix of Web site traffic between 197 referrers (rows) and 203 Web site pages (columns) is clustered into a 26×22 matrix with only 6% information loss. While KL distance is not actually a distance, because it is not symmetric,

	[1] /products/	[3] /company/jobs.html	[3] /products/hitlist.html	[3] /knowledgejstaycl0se5bl	[3] /insighthrbw-3 0.1/generid...	[1] /accrue3000c.html	[1] /company/	[2] /	[7] /services/subserv.html	[3] /index.html	[5] /insighthrbw-3 0.1/generid...	[1] /products/decision_ds.html	[1] /dx-cgi/product=insightver...	[2] /support/	[16] /pdf/wp_0599part1.pdf
[1] corporate-ir.NET								1	194						
[2] accrue.COM	5				30		8	152			73		5	26	
[20] harcourtonline.COM		1			71			43			242			37	
[3] issel.UK								1		27					
[3] itmcenter.COM								9							58
[1] about.COM	32		27												
[13] altavista.COM	481		1				68	644		1					
[3] google.COM	12	3	24				1	169		1		1			2
[12] excite.COM	5			3				473	6	35					1
[1] bayareacareers.COM															
[5] freelinks.COM	83							1							
[1] rulequest.COM												108			
[6] go.COM	192	5	13				28	460		58		1			2
[2] n/a.-	992	299	1018	141	154	149	153	38419	12559	2191	240	111	14	89	5362
[3] yahoo.COM	529	1		2				1148	1	9					4
[5] n/a.NUM	7		2	1	134			82	3	12	324		25	30	
[1] digitalmass.COM						236									
[14] bolt.COM					153			2			100		2	11	
[8] collegeclub.COM					10			27			16		4	23	
[8] autozone.COM					6			16			8		44	1	
[68] searchenginewatch.COM	4		1					2241		2					
[6] microsoft.COM	232	1	1	1			13	752	2	7					
[1] webtop.COM		1						1	5	1					6
[5] scripps.COM								2						59	
[1] rsvp0.NET				188						1					

Current View: Analysis Normal
Data Set Rows: 197
Data Set Cols: 203
Top Group Rows: 26
Top Group Cols: 22
Analysis Info Loss: 5.97%

Fig. 6. Algorithm Learning Relation

it can be symmetrized to the Jensen–Shannon divergence. Dhillon et al. [137] used Jensen–Shannon divergence to cluster words in k-means fashion in text classification. Besides text and Web mining, the idea of coclustering finds its way into other applications, as for example, clustering of gene microarrays [93].

10 General Algorithmic Issues

Clustering algorithms share some important common issues that need to be addressed to make them successful. Some issues are so ubiquitous that they are not even specific to unsupervised learning and can be considered as a part of an overall data mining framework. Other issues are resolved in certain algorithms we presented. In fact, many algorithms were specifically designed to address some of these issues:

- Assessment of results,
- Choice of appropriate number of clusters,
- Data preparation,
- Proximity measures,
- Handling outliers.

10.1 Assessment of Results

The data mining clustering process starts with the assessment of whether any cluster tendency can be established in data, and correspondingly includes some attribute selection, and in many cases, feature construction. Assessment finishes with the validation and evaluation of the resulting clustering system. Clusters can be assessed by an expert, or by a particular automated procedure. Traditionally, the first type of assessment relates to two issues: (1) cluster interpretability, (2) cluster visualization. Interpretability depends on the technique used. Model-based probabilistic and conceptual algorithms, such as COBWEB, have better scores in this regard. k-Means and k-medoid methods generate clusters that are interpreted as dense areas around centroids or medoids and, therefore, also score well. Jain et al. [245] cover cluster validation thoroughly. A discussion of cluster visualization and related references can be found in [254].

Regarding automatic procedures, when two partitions are constructed (with the same or different number of subsets k), the first order of business is to compare them. Sometimes the actual class label of one partition is known. The situation is similar to testing a classifier in predictive mining when the actual target is known. Comparison of s and j labels is similar to computing an error, confusion matrix, etc. in predictive mining. Simple Rand criterion serves this purpose [366]. Computation of a *Rand* index (defined below) involves pairs of points that were assigned to the same and to different clusters in each of two partitions. Hence it has $O\left(N^2\right)$ complexity and is not always feasible. *Conditional entropy* of a known label s given clustering partitioning [115],

$$H(S\,|\,J) = -\sum_{j} p_j \sum_{s} p_{s|j} \log\left(p_{s|j}\right),$$

is another measure used. Here $p_j, p_{s|j}$ are probabilities of j cluster, and conditional probabilities of s given j. It has $O\left(N\right)$ complexity. Other measures are also used, for example, the *F-measure* [303]. Meila [334] explores comparing clusters in detail.

10.2 How Many Clusters?

In many methods, the number k of clusters to construct is an input parameter. Larger k results in more granular and less-separated clusters. In the case of k-means, the objective function is monotone decreasing. Therefore, the question of how to choose k is not trivial.

Many criteria have been introduced to find an optimal k. Some industrial applications (e.g., SAS) report a pseudo F-statistics. This only makes sense for k-means clustering in the context of ANOVA. Earlier publications on the subject analyzed cluster separation for different k [147,336]. For instance, the distance between any two centroids (medoids) normalized by the corresponding cluster's radii (standard deviations) and averaged (with cluster weights) is a reasonable choice for the *coefficient of separation*. This coefficient has very low complexity. Another popular choice for a separation measure is the Silhouette coefficient [265] having $O\left(N^2\right)$ complexity, which is used in conjunction with CLARANS in [349]. Consider the average distance between point x of cluster C and other points within C and compare it with the corresponding average distance of the best fitting cluster G other than C

$$a(x) = \frac{1}{|C| - 1} \sum_{y \in C, y \neq x} d(x, y), \quad b(x) = \min_{G \neq C} \frac{1}{|G|} \sum_{y \in G} d(x, y).$$

The Silhouette coefficient of x is $s(x) = (b(x) - a(x))/max\{a(x), b(x)\}$, with values close to $+1$ corresponding to a good clustering choice and values below 0 to a bad clustering choice. The overall average of individual $s(x)$ gives a good indication of cluster system appropriateness.

Remember that assignment of a point to a particular cluster may often involves a certain arbitrariness. Depending on how well a point fits a cluster C, different probabilities or weights $w(x, C)$ can be introduced so that a hard (strict) assignment is defined as

$$C(x) = \mathrm{argmin}_C w(x, C).$$

A Partition coefficient [69] having $O(kN)$ complexity is equal to the sum of squares of the weights

$$W = \frac{1}{N} \sum_{x \in X} w(x, C(x))^2$$

(compare with GINI index). Each of the discussed measures can be plotted as a function of k and the graph can be inspected to help choose the best k.

A strong probabilistic foundation of the *mixture model*, discussed in Sect. 3.1, allows viewing a choice of optimal k as a problem of fitting the data by the best model. The question is whether adding new parameters results in a better model. In Bayesian learning (for example, AUTOCLASS [105]) the likelihood of the model is directly affected (through priors) by the model complexity (i.e., the number of parameters proportional to k). Several criteria were suggested including:

- Minimum description length (MDL) criterion [370, 371, 390];
- Minimum message length (MML) criterion [435, 436];
- Bayesian information criterion (BIC) [171, 390];
- Akaike's information criterion (AIC) [82];
- Noncoding information theoretic criterion (ICOMP) [83];
- Approximate weight of evidence (AWE) criterion [48];
- Bayes factors [264].

All these criteria are expressed through combinations of log-likelihood L, number of clusters k, number of parameters per cluster, total number of estimated parameters p, and different flavors of Fisher information matrix. For example,

$$\text{MDL}(k) = -L + p/2 - log(p), \quad k_{\text{best}} = \text{argmin}_k \text{MDL}(k),$$

$$\text{BIC}(k) = L - \frac{p}{2} \cdot log(n), \quad k_{\text{best}} = \text{argmax}_k \text{BIC}(k).$$

Further details and discussion can be found in [73, 171, 352]. Here are a few examples of criteria usage: MCLUST and X-means use the BIC criterion, SNOB uses the MML criterion, while CLIQUE and the evolutionary approach to k determination [305] use MDL. Significant expertise has been developed in estimating goodness of fit based on the criteria above. For example, different ranges of BIC are suggested for weak, positive, and very strong evidence in favor of one clustering system versus another in [172]. Smyth [402] suggested a likelihood crossvalidation technique for determining the best k.

10.3 Data Preparation

Irrelevant attributes make chances of a successful clustering futile, because they negatively affect proximity measures and eliminate clustering tendency. Therefore, sound exploratory data analysis (EDA) is essential. An overall framework for EDA can be found in [50]. As its first order of business, EDA eliminates inappropriate attributes and reduces the cardinality of the retained categorical attributes. Next it provides attribute selection. Different attribute selection methods exist. Inconsistency rates are utilized in [321]. The idea of a Markov blanket is used in [291]. While there are other methods (for example, [248]), most are used primarily for predictive and not descriptive mining and thus do not address general-purpose attribute selection for clustering. We conclude that cluster-specific attribute selection has yet to be invented.

Attribute transformation has already been discussed in the context of dimensionality reduction. The practice of assigning different weights to attributes and/or scaling of their values (especially, standardization) is widespread and allows constructing clusters of better shapes. To some extent *attribute scaling* can be viewed as the continuation of attribute selection.

In real-life applications it is crucial to handle attributes of different types. For example, images are characterized by color, texture, shape, and location, resulting in four attribute subsets. Modha and Spangler [341] suggested a very interesting approach for attribute scaling that pursues the objective of clustering in each attribute subset by computing weights (a simplex) that minimize the product of intracluster to intercluster ratios for the attribute subset projections (called the *generalized Fisher ratio*).

In many applications data points have different significance. For example, in assortment planning, stores can be characterized by the profiles of the percentage sales of particular items. However, the overall sale volume gives an additional weight to larger stores. Partitioning relocation and grid-based algorithms easily handle weighted data (centroids become centers of weights instead of means). This practice is called *case scaling*.

Some algorithms, for example, the extension of the algorithm CLARANS [153] and the algorithm DBSCAN [152], depend on the effectiveness of data access. To facilitate this process data indices are constructed. Index structures used for spatial data include KD-trees [177], R-trees [211], and R^*-trees [293]. A blend of attribute transformations (DFT, Polynomials) and indexing techniques is presented in [268]. Other indices and numerous generalizations exist [51, 57, 157, 259, 269, 438]. The major application of such data structures is the nearest neighbor search.

One way to preprocess multimedia data is to embed it into Euclidean space. In this regard see the algorithm FastMap [158].

A fairly diverse range of preprocessing is used for variable length sequences. Instead of handling them directly (as discussed in the Sect. 3.1), a fixed set of features representing the variable length sequences can be derived [210, 324].

10.4 Proximity Measures

Different distances and similarity measures are used in clustering [242]. The usual L_p distance

$$d(x, y) = \|x - y\|, \quad \|z\|_p = \left(\sum_{j=1:d} |z_j|^p \right)^{1/p}, \quad \|z\| = \|z\|_2$$

is used for numerical data, $1 \le p < \infty$; lower p corresponds to a more robust estimation (therefore, less affected by outliers). Euclidean ($p = 2$) distance is by far the most popular choice used in k-means objective functions (the sum of squares of distances between points and centroids) that has a clear statistical meaning as the total intercluster variance. Manhattan distance corresponds to $p = 1$. The distance that returns the maximum of absolute difference in coordinates is also used and corresponds to $p = \infty$. In many applications (e.g., profile analyses) points are scaled to have a unit norm, so that the proximity measure is an angle between the points,

$$d(x, y) = \arccos \left(x^{\mathrm{T}} \cdot y / \|x\| \cdot \|y\| \right).$$

This measure is used in specific tools, such as DIGNET [419], and in particular applications such as text mining [132]. All above distances assume attribute independence (diagonal covariance matrix S). The Mahanalabonis distance [326]

$$d(x,y) = \sqrt{(x-y)^T T S(x-y)}$$

is used in algorithms such as ORCLUS [13] that do not make this assumption.

Formula $s(x,y) = 1/(1 + d(x,y))$ defines similarity for whatever distance. For numerical attributes other choices include the *Cosine*, and *Dice* coefficients and distance *Eponent*

$$s_{cos} = x^T \cdot y/\|x\| \cdot \|y\|, \; s_{Dice} = 2x^T \cdot y/(\|x\| + \|y\|), \; s_{exp} = exp\left(-\|x-y\|^\alpha\right).$$

Now we shift our attention to categorical data. A number of similarity measures exist for categorical attributes [142, 155]. Assuming binary attributes with values $\alpha, \beta = \pm$, let $d_{\alpha\beta}$ be a number of attributes having outcomes α in x and β in y. Then the *Rand* and *Jaccard* (also known as *Tanimoto*) indices are defined as

$$R(x,y) = (d_{++} + d_{--})/(d_{++} + d_{+-} + d_{-+} + d_{--}),$$

$$J(x,y) = d_{++}/(d_{++} + d_{+-} + d_{-+}).$$

Notice that the Jaccard index treats positive and negative values asymmetrically, which makes it the measure of choice for transactional data, with a positive value meaning that an item is present. The Jaccard index is simply the fraction of common items of two transactions relative to the number of items in both transactions. It is also used in collaborative filtering, sequence analysis, text mining, and pattern recognition. The *Extended Jaccard* coefficient is advocated in [188]. For construction of similarity measures for market basket analysis see [11, 38]. Similarity can also be constructed axiomatically based on information-theoretical considerations [38, 316]. The last two references also contain material related to string similarity, where biology is one application. For strings over the same alphabet, *edit distance* is frequently used [25]. Edit distance is based on the length of a sequence of transformations (such as insertion, deletion, transposition, etc.) that are necessary to transform one string into another. A classic Hamming distance [115] is also used. Further references can be found in a review [245]. Historically, textual mining was a source of major inspirations for the concept of similarity [369].

10.5 Handling Outliers

Applications that derive their data from measurements have an associated amount of noise, which can be viewed as outliers. Alternately, outliers can be viewed as legitimate records having abnormal behavior. In general, clustering techniques do not distinguish between the two; neither noise nor abnormalities fit into clusters. Correspondingly, the preferred way to deal with outliers in partitioning the data is to keep outliers separate, and thereby avoid polluting actual clusters.

Descriptive learning handles outliers in many ways. If a summarization or a data preprocessing phase is present, it usually takes care of outliers. For example, this is the case with grid-based methods, which rely on input thresholds to eliminate low-populated cells. Algorithms described in Sect. 8 provide further examples. The algorithm BIRCH [459, 460] revisits outliers during the major CF tree rebuilds, but in general it handles them separately. This approach is shared by similar systems [111]. Framework suggested by Bradley et al. in [85] utilizes a multiphase approach to handle outliers.

Certain algorithms have specific features for outlier handling. The algorithm CURE [207] uses shrinkage of a cluster's representatives to suppress the effects of outliers. K-medoids methods are generally more robust than k-means methods with respect to outliers: medoids do not "feel" outliers (median vs. average). The algorithm DBCSAN [152] uses the concepts of internal (core), boundary (reachable), and outlier (nonreachable) points. The algorithm CLIQUE [15] goes a step further by eliminating subspaces with low coverage. The algorithm WaveCluster [394] is known to handle outliers very well because of its DSP roots. The algorithm ORCLUS [13] produces a partition plus a set of outliers.

What precisely is an outlier? Statistics defines an outlier as a point that does not fit a probability distribution. Classic data analysis utilizes a concept of *depth* [424] and defines an outlier as a point of low depth. This concept becomes computationally infeasible for $d > 3$. Data mining is gradually developing its own definitions.

Consider two positive parameters ϵ, δ. A point can be declared an outlier if its ϵ-neighborhood contains less than $1 - \delta$ fraction of a whole dataset X [273]. Ramaswamy et al. [365] noticed that this definition can be improved by eliminating the parameter ϵ. Rank all the points by their distance to the kth nearest neighbor and define the δ fraction of points with highest ranks as outliers. Both these definitions are uniformly global, so the question of how to describe local outliers remains. In essence, different subsets of data have different densities and may be governed by different distributions. A point close to a tight cluster might be more likely to be an outlier than a point that is further away from a more dispersed cluster. A concept of *local outlier factor* (LOF) that specifies a degree of outlierness is developed in [90]. The definition is based on the distance to the kth nearest neighbor. Knorr et al. [274] addressed a related problem of how to eliminate outliers in order to compute an appropriate covariance matrix that describes a given locality. To do so, they utilized the *Donoho–Stahel estimator* in two-dimensional space.

Crude handling of outliers works surprisingly well in many applications, because many applications are concerned with systematic patterns. An example is customer segmentation with the objective of finding a segment for a direct mail campaign. On the other hand, philosophically an outlier is an atypical leftover after regular clustering and, as such, it might have its own significance. Therefore, in addition to eliminating negative effects of outliers on cluster construction, there is a separate factor driving the interest in outlier detection, namely, that in some applications, the outlier is the commodity of trade. This is so in medical diagnostics, fraud detection, network security, anomaly detection, and computer immunology. Some connections and further references can be found in [167, 186, 307]. In CRM, E-commerce, and web site analytics outliers relate to the concepts of *interesting* and *unexpected* [355,356,360,397].

Acknowledgments

Many people either supported this work or helped me with the text. I thank them all. I am especially grateful to Jonathan Becher, Cliff Brunk, Usama Fayyad, Jiawei Han, Jacob Kogan, and Charles Nicholas. I am also thankful to my wife Rimma for continual encouragement over ups and downs of this work.

Similarity-Based Text Clustering: A Comparative Study

J. Ghosh and A. Strehl

Summary. Clustering of text documents enables unsupervised categorization and facilitates browsing and search. Any clustering method has to embed the objects to be clustered in a suitable representational space that provides a measure of (dis)similarity between any pair of objects. While several clustering methods and the associated similarity measures have been proposed in the past for text clustering, there is no systematic comparative study of the impact of similarity measures on the quality of document clusters, possibly because most popular cost criteria for evaluating cluster quality do not readily translate across qualitatively different measures. This chapter compares popular similarity measures (Euclidean, cosine, Pearson correlation, extended Jaccard) in conjunction with several clustering techniques (random, self-organizing feature map, hypergraph partitioning, generalized k-means, weighted graph partitioning), on a variety of high dimension sparse vector data sets representing text documents as bags of words. Performance is measured based on mutual information with a human-imposed classification. Our key findings are that in the quasiorthogonal space of word frequencies: (i) Cosine, correlation, and extended Jaccard similarities perform comparably; (ii) Euclidean distances do not work well; (iii) Graph partitioning tends to be superior especially when balanced clusters are desired; (iv) Performance curves generally do *not* cross.

1 Introduction

Document clusters can provide a structure for organizing large bodies of text for efficient browsing and searching. For example, recent advances in Internet search engines (e.g., www.vivisimo.com, www.metacrawler.com) exploit document cluster analysis. For this purpose, a document is commonly represented as a vector consisting of the suitably normalized frequency counts of words or terms. Each document typically contains only a small percentage of all the words ever used. If we consider each document as a multidimensional vector and then try to cluster documents based on their word contents, the problem differs from classic clustering scenarios in several ways: Document data are high dimensional[1], characterized by a very sparse term-document matrix with positive ordinal attribute values and a significant amount of outliers. In such situations, one is truly faced with the "curse of dimensionality" issue [176] since even after feature reduction, one is left with hundreds of dimensions per document.

Since clustering basically involves grouping objects based on their interrelationships or similarities, one can alternatively work in *similarity space* instead of the original feature space. The key insight is that if one can find a similarity measure (derived from the object features) that is appropriate for the problem domain, then a single number can capture the essential "closeness" of a given pair of objects, and any further analysis can be based only on these numbers. Once this is done, the original high-dimensional space is not dealt with at all; we only work in the transformed similarity space, and subsequent processing is independent of the dimensionality of the data [412]. A similar approach can be found in kernel based methods, such as Support Vector Machines (SVMs), for classification problems since the kernel function indicates a similarity measure obtained by a generalized inner product [240, 249, 430]. It is interesting to note that some very early works on clustering (e.g., [233]) were based on the concept of similarity, but subsequently the focus moved toward working with distances in a suitable embedding space, since typically, n, the number of objects considered, would be much larger than the number of features, d, used to represent each object. With text, d is very high; hence there is a renewal of interest in similarity-based approaches.

A typical pattern clustering activity involves the following five steps according to [242]:

1. Suitable object representation,
2. Definition of proximity between objects,
3. Clustering,
4. Data abstraction,
5. Assessment of output

The choice of similarity or distance in step 2 can have a profound impact on clustering quality. The significant amount of empirical studies in the 1980s and earlier on document clustering largely selected either Euclidean distance or cosine similarity, and emphasized various ways of representing/normalizing

---

[1]The dimension of a document in vector space representation is the size of the vocabulary, often in the tens of thousands.

documents before this step [377, 443]. Agglomerative clustering approaches were dominant and compared favorably with flat partitional approaches on small-sized or medium-sized collections [367, 443]. But lately, some new partitional methods have emerged (spherical k-means (KM), graph partitioning (GP) based, etc.) that have attractive properties in terms of both quality and scalability and can work with a wider range of similarity measures. In addition, much larger document collections are being generated.[2] This warrants an updated comparative study on text clustering, which is the motivation behind this chapter. Some very recent, notable comparative studies on document clustering [408, 463, 464] also consider some of these newer issues. Our work is distinguished from these efforts mainly by its focus on the key role of the similarity measures involved, emphasis on balancing, and the use of a normalized mutual information based evaluation that we believe has superior properties.

We mainly address steps 2 and 5 and also touch upon steps 3 and 4 in the document clustering domain. We first compare similarity measures analytically and illustrate their semantics geometrically (steps 2 and 4). Second, we propose an experimental methodology to compare high-dimensional clusterings based on mutual information and we argue why this is preferable to the more commonly used purity-based or entropy-based measures (step 5) [75, 408, 463]. Finally, we conduct a series of experiments to evaluate the performance and the cluster quality of four similarity measures (Euclidean, cosine, Pearson correlation, extended Jaccard) in combination with five algorithms (random, self-organizing map (SOM), hypergraph partitioning (HGP), generalized KM, weighted graph partitioning) (steps 2 and 3). Agglomerative clustering algorithms have been deliberately ignored even though they have been traditionally popular in the information retrieval community [367], but are not suitable for very large collections due to their computational complexity of at least $O(n^2 \log n)$ [300]. Indeed, if a hierarchy of documents is required, it is more practical to first partition the collection into an adequately large number (say 100 if finally about ten groups are desired) clusters, and then run an agglomerative algorithm on these partially summarized data.

Section 2 considers previous related work and Sect. 3 discusses various similarity measures.

2 Background and Notation

Clustering has been widely studied in several disciplines, especially since the early 1960s [59, 224, 243]. Some classic approaches include partitional methods such as k-means, hierarchical agglomerative clustering, unsupervised Bayes, and soft[3] techniques, such as those based on fuzzy logic or statistical mechanics

---

[2]IBM Patent Server has over 20 million patents. Lexis-Nexis contains over 1 billion documents

[3]In *soft* clustering, a record can belong to multiple clusters with different degrees of "association" [299].

[103]. Conceptual clustering [163], which maximizes category utility, a measure of predictability improvement in attribute values given a clustering, is also popular in the machine learning community. In most classical techniques, and even in fairly recent ones proposed in the data mining community (CLARANS, DBSCAN, BIRCH, CLIQUE, CURE, WAVECLUSTER, etc. [217, 368]) the objects to be clustered only have numerical attributes and are represented by low-dimensional feature vectors. The clustering algorithm is then based on distances between the samples in the original vector space [382]. Thus these techniques are faced with the "curse of dimensionality" and the associated sparsity issues, when dealing with very high-dimensional data such as text. Indeed, often, the performance of such clustering algorithms is demonstrated only on illustrative two-dimensional examples.

Clustering algorithms may take an alternative view based on a notion of *similarity* or dissimilarity. Similarity is often derived from the inner product between vector representations, a popular way to quantify document similarity. In [136], the authors present a spherical KM algorithm for document clustering using this similarity measure. Graph-based clustering approaches, which attempt to avoid the "curse of dimensionality" by transforming the problem formulation into a similarity space, include [75, 411, 461]. Finally, when only pairwise similarities are available, techniques such as Multi-Dimensional Scaling (MDS) [422] have been used to embed such points into a low-dimensional space such that the stress (relative difference between embedded point distances and actual distances) is minimized. Clustering can then be done in the embedding space. However, in document clustering this is not commonly used since for acceptable stress levels the dimensionality of the embedding space is too high.

Note that similarity-based methods take a *discriminative* approach to clustering. An alternative would be to take a *generative* viewpoint, starting from an underlying probabilistic model of the data and then finding suitable parameters typically through a maximum likelihood procedure. Cluster locations and properties are then derived as a by-product of this procedure. A detailed discussion of the pros and cons of discriminative approaches as compared to generative ones is given in [187]. Often discriminative approaches give better results, but any approach that required all-pairs similarity calculation is inherently at least $O(N^2)$ in both computational and storage requirements. In contrast, model-based generative approaches can be linear in N. A detailed empirical comparison of different model-based approaches to document clustering is available in [464] and hence we do not revisit these models in this chapter. Clustering has also been studied for the purpose of *browsing*. A two-dimensional SOM [284] has been applied to produce a map of, e.g., Usenet postings in WEBSOM [285]. The emphasis in WEBSOM is not to maximize cluster quality but to produce a human interpretable two-dimensional spatial map of known categories (e.g., newsgroups). In the Scatter/Gather approach [120] document clustering is used for improved interactive browsing of large query results. The focus on this work is mostly on speed/scalability and not necessary maximum cluster quality. In [451], the effectiveness of clustering for organizing web documents was studied.

There is also substantial work on *categorizing* documents. Here, since at least some of the documents have labels, a variety of supervised or semi-supervised techniques can be used [342, 350]. A technique using the support vector machine is discussed in [249]. There are several comparative studies on document classification [447, 448].

Dimensionality reduction for text classification/clustering has been studied as well. Often, the data are projected onto a small number of dimensions corresponding to principal components or a scalable approximation thereof (e.g., Fastmap [156]). In latent semantic indexing (LSI) [124] the term-document matrix is modeled by a rank-K approximation using the top K singular values. While LSI was originally used for improved query processing in information retrieval, the base idea can be employed for clustering as well.

In *bag-of-words* approaches the term-frequency matrix contains occurrence counts of terms in documents. Often, the matrix is preprocessed in order to enhance discrimination between documents. There are many schemes for selecting term, and global, normalization components. One popular preprocessing is normalized term frequency, inverse document frequency (TF-IDF), which also comes in several variants [40, 377]. However, this chapter does not discuss the properties of feature extraction, see, e.g., [312, 459] instead. In [447, 448] classification performance of several other preprocessing schemes is compared.

Following Occam's Razor, we do *not* use any weighting but use the raw frequency matrix of selected words for our comparison. Hence, appropriate normalization has to be encoded by the similarity measure.

Let n be the number of objects (documents, samples) in the data and d the number of features (words, terms) for each object $\mathbf{x}_j$ with $j \in \{1, \ldots, n\}$. Let k be the desired number of clusters. The input data can be represented by a $d \times n$ data matrix $\mathbf{X}$ with the jth column vector representing the sample $\mathbf{x}_j$. $\mathbf{x}_j^{\mathrm{T}}$ denotes the transpose of $\mathbf{x}_j$. Hard clustering assigns a label λ_j to each d-dimensional sample $\mathbf{x}_j$, such that similar samples tend to get the same label. In general the labels are treated as nominals with no inherent order, though in some cases, such as one-dimensional SOM or GP approaches based on swapping of vertices with neighboring partitions the labeling contains extra ordering information. Let $\mathcal{C}_\ell$ denote the set of all objects in the ℓth cluster $(\ell \in \{1, \ldots, k\})$, with $\mathbf{x}_j \in \mathcal{C}_\ell \Leftrightarrow \lambda_j = \ell$ and $n_\ell = |\mathcal{C}_\ell|$. The number of distinct labels is k, the desired number of clusters. We treat the labels as nominals with no order, though in some cases, such as the SOM or graph partitioning, the labeling may contain extra ordering information. Batch clustering proceeds from a set of raw object descriptions $\mathcal{X}$ via the vector space description $\mathbf{X}$ to the cluster labels λ ($\mathcal{X} \to \mathbf{X} \to \lambda$). Section 3 briefly describes the compared similarity measures.

3 Similarity Measures

In this section, we introduce several similarity measures, illustrate some of their properties, and show why we are interested in some but not others. In Sect. 4, the algorithms using these similarity measures are discussed.

3.1 Conversion from a Distance Metric

The Minkowski distances $L_p(\mathbf{x}_a, \mathbf{x}_b) = \left(\sum_{i=1}^{d} |\mathbf{x}_{i,a} - \mathbf{x}_{i,b}|^p \right)^{1/p}$ are commonly used when objects are represented in a vector space. For $p = 2$ we obtain the Euclidean distance. There are several possibilities for converting such a distance metric (in $[0, \inf)$) into a similarity measure (in $[0, 1]$; usually similarity of 1 corresponds to a distance of 0) by a monotonic decreasing function. For Euclidean space, a good choice is: similarity $= \exp(-(\text{distance})^2)$, as it relates the squared error loss function to the negative log-likelihood for a Gaussian model for each cluster. In this chapter, we use the Euclidean $[0, 1]$-normalized similarity expressed by

$$s^{(\mathrm{E})}(\mathbf{x}_a, \mathbf{x}_b) = \mathrm{e}^{-\|\mathbf{x}_a - \mathbf{x}_b\|_2^2} \tag{1}$$

rather than alternatives such as $s(\mathbf{x}_a, \mathbf{x}_b) = 1/(1 + \|\mathbf{x}_a - \mathbf{x}_b\|_2)$.

3.2 Cosine Measure

A popular measure of similarity for text clustering is the cosine of the angle between two vectors. The cosine measure is given by

$$s^{(\mathrm{C})}(\mathbf{x}_a, \mathbf{x}_b) = \frac{\mathbf{x}_a^{\mathrm{T}} \mathbf{x}_b}{\|\mathbf{x}_a\|_2 \cdot \|\mathbf{x}_b\|_2} \tag{2}$$

and captures a scale invariant understanding of similarity. The cosine similarity does not depend on the length of the vectors, only their direction. This allows documents with the same relative distribution of terms to be treated identically. Being insensitive to the size of the documents makes this a very popular measure for text documents. Also, due to this property, document vectors can be normalized to the unit sphere for more efficient processing [136].

3.3 Pearson Correlation

In collaborative filtering, correlation is often used to predict a feature from a highly similar mentor group of objects whose features are known. The $[0, 1]$ normalized Pearson correlation is defined as

$$s^{(\mathrm{P})}(\mathbf{x}_a, \mathbf{x}_b) = \frac{1}{2} \left(\frac{(\mathbf{x}_a - \bar{x}_a)^{\mathrm{T}}(\mathbf{x}_b - \bar{x}_b)}{\|\mathbf{x}_a - \bar{x}_a\|_2 \cdot \|\mathbf{x}_b - \bar{x}_b\|_2} + 1 \right), \tag{3}$$

where $\bar{x}$ denotes the average feature values of $\mathbf{x}$. Note that this definition of Pearson correlation tends to give a full matrix. Other important correlations have been proposed, such as Spearman correlation [406], which works well on rank orders.

3.4 Extended Jaccard Similarity

The binary Jaccard coefficient[4] measures the degree of overlap between two sets and is computed as the ratio of the number of shared attributes (words) of $\mathbf{x}_a$ AND $\mathbf{x}_b$ to the number possessed by $\mathbf{x}_a$ OR $\mathbf{x}_b$. For example, given two sets' binary indicator vectors $\mathbf{x}_a = (0, 1, 1, 0)^T$ and $\mathbf{x}_b = (1, 1, 0, 0)^T$, the cardinality of their intersect is 1 and the cardinality of their union is 3, rendering their Jaccard coefficient $1/3$. The binary Jaccard coefficient is often used in retail market-basket applications. The binary definition of Jaccard coefficient can be extended to continuous or discrete non-negative features as:

$$s^{(J)}(\mathbf{x}_a, \mathbf{x}_b) = \frac{\mathbf{x}_a^T \mathbf{x}_b}{\|\mathbf{x}_a\|_2^2 + \|\mathbf{x}_b\|_2^2 - \mathbf{x}_a^T \mathbf{x}_b}, \tag{4}$$

which is equivalent to the binary version when the feature vector entries are binary. Extended Jaccard similarity retains the sparsity property of the cosine while allowing discrimination of collinear vectors as we show in Sect. 3.6. Another similarity measure highly related to the extended Jaccard is the Dice coefficient

$$s^{(D)}(\mathbf{x}_a, \mathbf{x}_b) = \frac{2\mathbf{x}_a^T \mathbf{x}_b}{\|\mathbf{x}_a\|_2^2 + \|\mathbf{x}_b\|_2^2}.$$

The Dice coefficient can be obtained from the extended Jaccard coefficient by adding $\mathbf{x}_a^T \mathbf{x}_b$ to both the numerator and the denominator. It is omitted here since it behaves very similar to the extended Jaccard coefficient.

3.5 Other (Dis-)Similarity Measures

Many other (dis-)similarity measures, such as shared nearest neighbor [247] or the edit distance, are possible [243]. In fact, the ugly duckling theorem states [442] the somewhat "unintuitive" fact that there is no way to distinguish between two different classes of objects, when they are compared over all possible features. As a consequence, any two arbitrary objects are equally similar unless we use domain knowledge. The similarity measures discussed in Sects. 3.1–3.4 are some of the popular ones that have been previously applied to text documents [170, 377].

---

[4] Also called the Tanimoto coefficient in the vision community.

3.6 Discussion

Clearly, if clusters are to be meaningful, the similarity measure should be invariant to transformations natural to the problem domain. Also, normalization may strongly affect clustering in a positive or a negative way. The features have to be chosen carefully to be on comparable scales and similarity has to reflect the underlying semantics for the given task.

Euclidean similarity is translation invariant but scale sensitive while cosine is translation sensitive but scale invariant. The extended Jaccard has aspects of both properties as illustrated in Fig. 1. Iso-similarity lines at $s = 0.25, 0.5$, and 0.75 for points $\mathbf{x}_1 = (3, 1)^\mathrm{T}$ and $\mathbf{x}_2 = (1, 2)^\mathrm{T}$ are shown for Euclidean, cosine, and the extended Jaccard. For cosine similarity only the 4 (out of 12) lines that are in the positive quadrant are plotted: The two lines in the lower right part are one of two lines from $\mathbf{x}_1$ at 0.5 and 0.75. The two lines in the upper left are for $\mathbf{x}_2$ at $s = 0.5$ and 0.75. The dashed line marks the locus of equal similarity to $\mathbf{x}_1$ and $\mathbf{x}_2$, which always passes through the origin for cosine and the extended Jaccard similarity.

Using Euclidean similarity $s^{(\mathrm{E})}$, isosimilarities are concentric hyperspheres around the considered point. Due to the finite range of similarity, the radius decreases hyperbolically as $s^{(\mathrm{E})}$ increases linearly. The radius does not depend on the center point. The only location with similarity of 1 is the considered point itself and all finite locations have a similarity greater than 0. This last property tends to generate nonsparse similarity matrices. Using the cosine measure $s^{(\mathrm{C})}$ renders the isosimilarities to be hypercones all having their apex at the origin and the axis aligned with the considered point. Locations with similarity 1 are on the one-dimensional subspace defined by this axis. The locus of points with similarity 0 is the hyperplane through the origin and perpendicular to this axis. For the extended Jaccard similarity $s^{(\mathrm{J})}$, the isosimilarities are nonconcentric hyperspheres. The only location with similarity 1

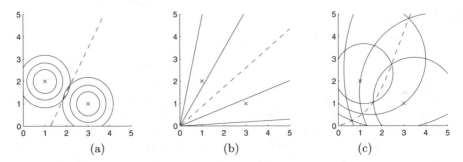

(a)   (b)   (c)

Fig. 1. Properties of **(a)** Euclidean-based, **(b)** cosine, and **(c)** extended Jaccard similarity measures illustrated in two dimensions. Two points $(1, 2)^\mathrm{T}$ and $(3, 1)^\mathrm{T}$ are marked with $\times$. For each point isosimilarity surfaces for $s = 0.25, 0.5$, and 0.75 are shown with solid lines. The surface that is equisimilar to the two points is marked with a dashed line

is the point itself. The hypersphere radius increases with the distance of the considered point from the origin so that longer vectors turn out to be more tolerant in terms of similarity than smaller vectors. Sphere radius also increases with similarity, and as $s^{(J)}$ approaches 0 the radius becomes infinite, rendering the sphere to the same hyperplane as obtained for cosine similarity. Thus, for $s^{(J)} \to 0$, the extended Jaccard behaves like the cosine measure, and for $s^{(J)} \to 1$, it behaves like the Euclidean distance.

In traditional Euclidean k-means clustering, the optimal cluster representative $\mathbf{c}_\ell$ minimizes the sum of squared error criterion, i.e.,

$$\mathbf{c}_\ell = \arg\min_{\mathbf{z} \in \mathcal{F}} \sum_{\mathbf{x}_j \in \mathcal{C}_\ell} \|\mathbf{x}_j - \mathbf{z}\|_2^2. \tag{5}$$

In the following, we show how this convex distance-based objective can be translated and extended to similarity space. Consider the generalized objective function $f(\mathcal{C}_\ell, \mathbf{z})$ given a cluster $\mathcal{C}_\ell$ and a representative $\mathbf{z}$:

$$f(\mathcal{C}_\ell, \mathbf{z}) = \sum_{\mathbf{x}_j \in \mathcal{C}_\ell} d(\mathbf{x}_j, \mathbf{z})^2 = \sum_{\mathbf{x}_j \in \mathcal{C}_\ell} \|\mathbf{x}_j - \mathbf{z}\|_2^2. \tag{6}$$

We use the transformation from (1) to express the objective in terms of similarity rather than distance:

$$f(\mathcal{C}_\ell, \mathbf{z}) = \sum_{\mathbf{x}_j \in \mathcal{C}_\ell} -\log(s(\mathbf{x}_j, \mathbf{z})). \tag{7}$$

Finally, we simplify and transform the objective using a strictly monotonic decreasing function: Instead of minimizing $f(\mathcal{C}_\ell, \mathbf{z})$, we maximize $f'(\mathcal{C}_\ell, \mathbf{z}) = e^{-f(\mathcal{C}_\ell, \mathbf{z})}$. Thus, in similarity space, the least squared error representative $\mathbf{c}_\ell \in \mathcal{F}$ for a cluster $\mathcal{C}_\ell$ satisfies

$$\mathbf{c}_\ell = \arg\max_{\mathbf{z} \in \mathcal{F}} \prod_{\mathbf{x}_j \in \mathcal{C}_\ell} s(\mathbf{x}_j, \mathbf{z}). \tag{8}$$

Using the concave evaluation function f', we can obtain optimal representatives for non-Euclidean similarity spaces.

To illustrate the values of the evaluation function $f'(\{\mathbf{x}_1, \mathbf{x}_2\}, \mathbf{z})$ are used to shade the background in Fig. 2. The maximum likelihood representative of $\mathbf{x}_1$ and $\mathbf{x}_2$ is marked with an $*$ in Fig. 2. For cosine similarity all points on the equi-similarity are optimal representatives. In a maximum likelihood interpretation, we constructed the distance similarity transformation such that $p(\mathbf{z}|\mathbf{c}_\ell) \sim s(\mathbf{z}, \mathbf{c}_\ell)$. Consequently, we can use the dual interpretations of probabilities in similarity space and errors in distance space.

4 Algorithms

In this section, we briefly summarize the algorithms used in our comparison. A random algorithm is used as a baseline to compare the result quality of KM, GP, HGP, and SOM.

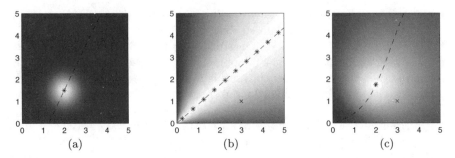

Fig. 2. More similarity properties shown on the two-dimensional example of Fig. 1. The goodness of a location as the common representative of the two points is indicated with brightness. The best representative is marked with an ∗. **(c)** The extended Jaccard adopts the middle ground between **(a)** Euclidean and **(b)** cosine-based similarity

4.1 Random Baseline

As a baseline for comparing algorithms, we use clustering labels drawn from a uniform random distribution over the integers from 1 to k. The complexity of this algorithm is $O(n)$.

4.2 Weighted Graph Partitioning

Clustering can be posed as a GP problem. The objects are viewed as the set of vertices $\mathcal{V}$. Two documents $\mathbf{x}_a$ and $\mathbf{x}_b$ (or vertices v_a and v_b) are connected with an undirected edge of positive weight $s(\mathbf{x}_a, \mathbf{x}_b)$, or $(a, b, s(\mathbf{x}_a, \mathbf{x}_b)) \in \mathcal{E}$. The cardinality of the set of edges $|\mathcal{E}|$ equals the number of *nonzero* similarities between all pairs of samples. A set of edges whose removal partitions a graph $\mathcal{G} = (\mathcal{V}, \mathcal{E})$ into k pairwise disjoint subgraphs $\mathcal{G}_\ell = (\mathcal{V}_\ell, \mathcal{E}_\ell)$ is called an edge separator. The objective in GP is to find such a separator with a minimum sum of edge weights. While striving for the minimum cut objective, the number of objects in each cluster has to be kept approximately equal. We produce balanced (equal sized) clusters from the similarity matrix using the multilevel graph partitioner Metis [262]. The most expensive step in this $O(n^2 \cdot d)$ technique is the computation of the $n \times n$ similarity matrix. In document clustering, sparsity can be induced by looking only at the v strongest edges or at the subgraph induced by pruning all edges except the v nearest neighbors for each vertex. Sparsity makes this approach feasible for large data sets. Sparsity is induced by particular similarities definitions based, for example, on the cosine of document vectors.

4.3 Hypergraph Partitioning

A hypergraph is a graph whose edges can connect more than two vertices (hyperedges). The clustering problem is then formulated as a finding the minimum cut of a hypergraph. A minimum cut is the removal of the set of hyperedges (with minimum edge weight) that separates the hypergraph into k unconnected components. Again, an object $\mathbf{x}_j$ maps to a vertex v_j. Each word (feature) maps to a hyperedge connecting all vertices with nonzero frequency count of this word. The weight of this hyperedge is chosen to be the total number of occurrences in the data set. Hence, the importance of a hyperedge during partitioning is proportional to the occurrence of the corresponding word. The minimum cut of this hypergraph into k unconnected components gives the desired clustering. We employ the hMetis package [263] for partitioning. An advantage of this approach is that the clustering problem can be mapped to a graph problem without the explicit computation of similarity, which makes this approach computationally efficient with $O(n \cdot d \cdot k)$ assuming a (close to) linear performing hypergraph partitioner. Note that samplewise frequency information gets lost in this formulation since there is only a single weight associated with a hyperedge.

4.4 Self-organizing Map

The SOM [70, 284] is a popular topology preserving clustering algorithm with nice visualization properties. For simplicity, we only use a one-dimensional line topology. Also, two-dimensional or higher dimensional topologies can be used. To generate k clusters we use k cells in a line topology and train the network for $m = 5,000$ epochs or 10 min (whichever comes first). All experiments are run on a dual processor 450 MHz Pentium using the SOM implementation in the Matlab neural network toolbox. The resulting network is subsequently used to generate the label vector λ from the index of the most activated neuron for each sample. The complexity of this incremental algorithm is $O(n \cdot d \cdot k \cdot m)$ and mostly determined by the number of epochs m and samples n.

4.5 Generalized k-means

The KM algorithm using the squared Euclidean or Mahalonobis distances as a measure of divergence, is perhaps the most popular partitional approach to clustering. This is really a generative approach, being a limiting case of soft clustering achieved by fitting a mixture of Gaussians to the data via the EM algorithm [266]. It has been recently shown that the scope of this framework is very broad, the essential properties of KM carry over to all regular Bregman divergences (and only to this class of divergence measures), and a similar generalization is also possible for the soft version [46]. The complexity of this set of algorithms is $O(n \cdot d \cdot k \cdot m)$, where m is the number of iterations needed for convergence.

Given the popularity of KM, we decided to convert cosine, Jaccard, and Pearson similarity measures into the corresponding divergences using (1), in addition to retaining the squared Euclidean distance to obtain four versions of KM. However we have not considered the use of KL-divergence, which has a natural correspondence with multinomial mixture modeling, as extensive work using this information theoretic measure is already available [463].

4.6 Other Clustering Methods

Several other clustering methods have also been considered but have not been used in our experimental comparison. Agglomerative models (single link, average link, complete link) [143] are computationally expensive (at least $O(n^2 \log n)$) and often result in highly skewed trees, indicating domination by one very large cluster. A detailed comparative study of generative, mixture model-based approaches to text, is available from [464]. Certain clustering algorithms from the data mining community (e.g., CLARANS, DBSCAN, BIRCH, CLIQUE, CURE, WAVECLUSTER [217, 368]) have been omitted since they are mostly scalable versions designed for low-dimensional data. Partitioning approaches based on principal directions have not been shown here since they perform comparably to hierachical agglomerative clustering [75]. Other GP approaches such as spectral bisectioning [227] are not included since they are already represented by the multilevel partitioner Metis.

5 Evaluation Methodology

We conducted experiments with all five algorithms, using four variants (involving different similarity measures) each for KM and GP, yielding 11 techniques in total. This section gives an overview of ways to evaluate clustering results. A good recent survey on clustering evaluation can be found in [463], where the emphasis is on determining the impact of a variety of cost functions, built using distance or cosine similarity measures, on the quality of two generic clustering approaches.

There are two fundamentally different ways of evaluating the quality of results delivered by a clustering algorithm. *Internal* criteria formulate quality as a function of the given data and/or similarities. For example, the mean squared error criterion is a popular evaluation criterion. Hence, the clusterer can evaluate its own performance and tune its results accordingly. When using internal criteria, clustering becomes an optimization problem. *External* criteria impose quality by additional, external information not given to the clusterer, such as class labels. While this makes the problem ill-defined, it is sometimes more appropriate since groupings are ultimately evaluated externally by humans.

5.1 Internal (Model-Based, Unsupervised) Quality

Internal quality measures, such as the sum of squared errors, have traditionally been used extensively. Given an internal quality measure, clustering can be posed as an optimization problem that is typically solved via greedy search. For example, KM has been shown to greedily optimize the sum of squared errors.

- Error (mean/sum-of-squared error, scatter matrices)
  The most popular cost function is the scatter of the points in each cluster. Cost is measured as the mean square error of data points compared to their respective cluster centroid. The well-known KM algorithm has been shown to heuristically minimize the squared error objective. Let n_ℓ be the number of objects in cluster $\mathcal{C}_\ell$ according to λ. Then, the cluster centroids are

$$c_\ell = \frac{1}{n_\ell} \sum_{\lambda_j = \ell} \mathbf{x}_j. \tag{9}$$

The sum of squared errors (SSE) is

$$\text{SSE}(\mathbf{X}, \lambda) == \sum_{\ell=1}^{k} \sum_{\mathbf{x} \in \mathcal{C}_\ell} \|\mathbf{x} - \mathbf{c}_\ell\|_2^2. \tag{10}$$

Note that the SSE formulation can be extended to other similarities by using $\text{SSE}(\mathbf{X}, \lambda) = \sum_{\ell=1}^{k} \sum_{\mathbf{x} \in \mathcal{C}_\ell} - \log s(\mathbf{x}, \mathbf{c}_\ell)$. Since we are interested in a quality measure ranging from 0 to 1, where 1 indicates a perfect clustering, we define quality as

$$\phi^{(S)}(\mathbf{X}, \lambda) = e^{-\text{SSE}(\mathbf{X}, \lambda)}. \tag{11}$$

This objective can also be viewed from a probability density estimation perspective using EM [126]. Assuming the data are generated by a mixture of multivariate Gaussians with identical, diagonal covariance matrices, the SSE objective is equivalent to maximizing the likelihood of observing the data by adjusting the centers and minimizing weights of the Gaussian mixture.

- Edge cut
  When clustering is posed as a GP problem, the objective is to minimize edge cut. Formulated as a $[0, 1]$-quality maximization problem, the objective is the ratio of remaining edge weights to total precut edge weights:

$$\phi^{(C)}(\mathbf{X}, \lambda) = \frac{\sum_{\ell=1}^{k} \sum_{a \in \mathcal{C}_\ell} \sum_{b \in \mathcal{C}_\ell, b > a} s(\mathbf{x}_a, \mathbf{x}_b)}{\sum_{a=1}^{n} \sum_{b=a+1}^{n} s(\mathbf{x}_a, \mathbf{x}_b)} \tag{12}$$

Note that this quality measure can be trivially maximized when there are no restrictions on the sizes of clusters. In other words, edge cut quality evaluation is only fair when the compared clusterings are well balanced. Let us define the balance of a clustering λ as

$$\phi^{(\text{BAL})}(\lambda) = \frac{n/k}{\max_{\ell \in \{1, \dots, k\}} n_\ell}. \tag{13}$$

A balance of 1 indicates that all clusters have the same size. In certain applications, balanced clusters are desirable because each cluster represents an equally important share of the data. Balancing has application-driven advantages, e.g., for distribution, navigation, summarization of the clustered objects. In [409] retail customer clusters are balanced, so they represent an equal share of revenue. Balanced clustering for browsing text documents has also been proposed [44]. However, some natural classes may not be of equal size, so relaxed balancing may become necessary. A middle ground between no constraints on balancing (e.g., k-means) and tight balancing (e.g., GP) can be achieved by overclustering using a balanced algorithm and then merging clusters subsequently [461]

- Category Utility [162, 193]
  The category utility function measures quality as the increase in predictability of attributes given a clustering. Category utility is defined as the *increase* in the expected number of attribute values that can be correctly guessed given a partitioning, *over* the expected number of correct guesses with no such knowledge. A weighted average over categories allows comparison of different sized partitions. Recently, it has been shown that category utility is related to squared error criterion for a particular standard encoding [338], whose formulation is used here. For binary features (i.e., attributes) the probability of the ith attribute being 1 is the mean of the ith row of the data matrix $\mathbf{X}$:

$$\bar{x}_i = \frac{1}{n} \sum_{j=1}^{n} x_{i,j}. \tag{14}$$

The conditional probability of the ith attribute to be 1 given that the data point is in cluster ℓ is

$$\bar{x}_{i,\ell} = \frac{1}{n_\ell} \sum_{\lambda_j = \ell} x_{i,j}. \tag{15}$$

Hence, category utility can be written as

$$\phi^{(\mathrm{CU})}(\mathbf{X}, \lambda) = \frac{4}{d} \sum_{\ell=1}^{k} \frac{n_\ell}{n} \left[\left(\sum_{i=1}^{d} \left(\bar{x}_{i,\ell}^2 - \bar{x}_{i,\ell} \right) \right) - \left(\sum_{i=1}^{d} \left(\bar{x}_i^2 - \bar{x}_i \right) \right) \right]. \tag{16}$$

Note that this definition divides the standard category by d so that $\phi^{(\mathrm{CU})}$ never exceeds 1. Category utility is defined to maximize predictability of attributes for a clustering. This limits the scope of this quality measure to low-dimensional clustering problems (preferably with each dimension being a categorical variable with small cardinality). In high-dimensional problems, such as text clustering, the objective is *not* to be able to predict the appearance of any possible word in a document from a particular cluster. In fact, there might be more unique words/terms/phrases than documents in a small data set. In preliminary experiments, category utility did not succeed in differentiating among the compared approaches (including random partitioning).

Using internal quality measures, fair comparisons can only be made amongst clusterings with the same choices of vector representation and similarity/distance measure. For example, using edge cut in cosine-based similarity would not be meaningful for an evaluation of Euclidean KM. So, in many applications a consensus on the internal quality measure for clustering is not found. However, in situations where the pages are categorized (labeled) by an external source, there is a plausible way out!

5.2 External (Model-Free, Semisupervised) Quality

External quality measures require an external grouping, for example as indicated by category labels, that is assumed to be "correct." However, unlike in classification such ground truth is not available to the clustering algorithm. This class of evaluation measures can be used to compare start-to-end performance of any kind of clustering regardless of the models or the similarities used. However, since clustering is an unsupervised problem, the performance cannot be judged with the same certitude as for a classification problem. The external categorization might not be optimal at all. For example, the way Web pages are organized in the Yahoo! taxonomy is certainly not the best structure possible. However, achieving a grouping similar to the Yahoo! taxonomy is certainly indicative of successful clustering.

Given g categories (or classes) $\mathcal{K}_h$ ($h \in \{1, \ldots, g\}$), we denote the categorization label vector κ, where $x_a \in \mathcal{K}_h \Leftrightarrow \kappa_a = h$. Let $n^{(h)}$ be the number of objects in category $\mathcal{K}_h$ according to κ, and n_ℓ the number of objects in cluster $\mathcal{C}_\ell$ according to λ. Let $n_\ell^{(h)}$ denote the number of objects that are in cluster ℓ according to λ as well as in category h given by κ. There are several ways of comparing the class labels with cluster labels.

- Purity
  Purity can be interpreted as classification accuracy under the assumption that all objects of a cluster are classified to be members of the dominant class for that cluster. For a single cluster, $\mathcal{C}_\ell$, purity is defined as the ratio of the number of objects in the *dominant* category to the total number of objects:

$$\phi^{(\mathrm{A})}(\mathcal{C}_\ell, \kappa) = \frac{1}{n_\ell} \max_h (n_\ell^{(h)}). \tag{17}$$

To evaluate an entire clustering, one computes the average of the clusterwise purities weighted by cluster size:

$$\phi^{(\mathrm{A})}(\lambda, \kappa) = \frac{1}{n} \sum_{\ell=1}^{k} \max_h (n_\ell^{(h)}). \tag{18}$$

- Entropy [115]

  Entropy is a more comprehensive measure than purity since rather than just considering the number of objects "in" and "not in" the dominant class, it takes the entire distribution into account. Since a cluster with all objects from the same category has an entropy of 0, we define entropy-based quality as 1 minus the [0,1]-normalized entropy. We define entropy-based quality for each cluster as:

$$\phi^{(E)}(\mathcal{C}_\ell, \kappa) = 1 - \sum_{h=1}^{g} -\frac{n_\ell^{(h)}}{n_\ell} \log_g \left(\frac{n_\ell^{(h)}}{n_\ell} \right). \tag{19}$$

And through weighted averaging, the total entropy quality measure falls out to be:

$$\phi^{(E)}(\lambda, \kappa) = 1 + \frac{1}{n} \sum_{\ell=1}^{k} \sum_{h=1}^{g} n_\ell^{(h)} \log_g \left(\frac{n_\ell^{(h)}}{n_\ell} \right). \tag{20}$$

Both purity and entropy are biased to favor a large number of clusters. In fact, for both these criteria, the globally optimal value is trivially reached when each cluster is a single object!

- Precision, recall, and F-measure [429]

  Precision and recall are standard measures in the information retrieval community. If a cluster is viewed as the results of a query for a particular category, then precision is the fraction of correctly retrieved objects:

$$\phi^{(P)}(\mathcal{C}_\ell, \mathcal{K}_h) = n_\ell^{(h)}/n_\ell. \tag{21}$$

Recall is the fraction of correctly retrieved objects out of all matching objects in the database:

$$\phi^{(R)}(\mathcal{C}_\ell, \mathcal{K}_h) = n_\ell^{(h)}/n^{(h)}. \tag{22}$$

The F-measure combines precision and recall into a single number given a weighting factor. The F_1-measure combines precision and recall with equal weights. The following equation gives the F_1-measure when querying for a particular category $\mathcal{K}_h$

$$\phi^{(F_1)}(\mathcal{K}_h) = \max_\ell \frac{2 \, \phi^{(P)}(\mathcal{C}_\ell, \mathcal{K}_h) \, \phi^{(R)}(\mathcal{C}_\ell, \mathcal{K}_h)}{\phi^{(P)}(\mathcal{C}_\ell, \mathcal{K}_h) + \phi^{(R)}(\mathcal{C}_\ell, \mathcal{K}_h)} = \max_\ell \frac{2n_\ell^{(h)}}{n_\ell + n^{(h)}}. \tag{23}$$

Hence, for the entire clustering the total F_1-measure is:

$$\phi^{(F_1)}(\lambda, \kappa) = \frac{1}{n} \sum_{h=1}^{g} n^{(h)} \phi^{(F)}(\mathcal{K}_h) = \frac{1}{n} \sum_{h=1}^{g} n^{(h)} \max_\ell \frac{2n_\ell^{(h)}}{n_\ell + n^{(h)}}. \tag{24}$$

Unlike purity and entropy, the F_1-measure is not biased toward a larger number of clusters. In fact, it favors coarser clusterings. Another issue is that random clustering tends not to be evaluated at 0.

- Mutual information [115]
  Mutual information is the most theoretically well founded among the considered external quality measures [140]. It is symmetric in terms of κ and λ. Let X and Y be the random variables described by the cluster labeling λ and category labeling κ, respectively. Let $H(X)$ denote the entropy of a random variable X. Mutual information between two random variables is defined as

$$I(X,Y) = H(X) + H(Y) - H(X,Y). \tag{25}$$

Also,

$$I(X,Y) \leq \min(H(X), H(Y)). \tag{26}$$

Since $\min(H(X), H(Y)) \leq (H(X) + H(Y))/2$, a tight upper bound on $I(X,Y)$ is given by $(H(X) + H(Y))/2$. Thus, a worst-case upper bound for all possible labelings and categorizations is given by

$$I(X,Y) \leq \max_{X,Y} \left(\frac{H(X) + H(Y)}{2} \right). \tag{27}$$

Hence, we define [0,1]-normalized mutual information-based quality as

$$NI(X,Y) = \frac{2 \cdot I(X,Y)}{\max_X(H(X)) + \max_Y(H(Y))}. \tag{28}$$

Using

$$I(X,Y) = \sum_{x \in X} \sum_{y \in Y} p(x,y) \log \frac{p(x,y)}{p(x) \cdot p(y)}. \tag{29}$$

Note that normalizing by the geometric mean of $H(X)$ and $H(Y)$ instead of the arithmetic mean will also work [410].

Now, approximating probabilities with frequency counts yields our quality measure $\phi^{(\text{NMI})}$:

$$\phi^{(\text{NMI})}(\lambda, \kappa) = \frac{2 \cdot \sum_{\ell=1}^{k} \sum_{h=1}^{g} \frac{n_\ell^{(h)}}{n} \log \frac{n_\ell^{(h)}/n}{n^{(h)}/n \, n_\ell/n}}{\log(k) + \log(g)} \tag{30}$$

Basic simplifications yield:

$$\phi^{(\text{NMI})}(\lambda, \kappa) = \frac{2}{n} \sum_{\ell=1}^{k} \sum_{h=1}^{g} n_\ell^{(h)} \log_{k \cdot g} \left(\frac{n_\ell^{(h)} n}{n^{(h)} n_\ell} \right) \tag{31}$$

Mutual information is less prone to biases than purity, entropy, and the F_1-measure. Singletons are not evaluated as perfect. Random clustering has mutual information of 0 in the limit. However, the best possible labeling evaluates to less than 1, unless classes are balanced, i.e., of equal size.

Note that our normalization penalizes over-refinements unlike the standard mutual information.[5]

External criteria enable us to compare different clustering methods fairly provided the external ground truth is of good quality. One could argue against external criteria that clustering does not have to perform as well as classification. However, in many cases clustering is an interim step to better understand and characterize a complex data set before further analysis and modeling.

Normalized mutual information is our preferred choice of evaluation in Sect. 6, because it is a relatively unbiased measure for the usefulness of the knowledge captured in the clustering in predicting category labels. Another promising evaluation method based on PAC-MDL bounds is given in [45].

6 Experiments

6.1 Data Sets and Preprocessing

We chose four text data sets for comparison. Here we briefly describe them:

- YAH. These data were parsed from Yahoo! news web pages [75]. The 20 original categories for the pages are Business, Entertainment (no sub-category, art, cable, culture, film, industry, media, multimedia, music, online, people, review, stage, television, variety), Health, Politics, Sports, Technology. The data can be downloaded from ftp://ftp.cs.umn.edu/ dept/users/boley/ (K1 series).
- N20. The data contain roughly 1,000 postings each from the following 20 newsgroup topics [302][6]:
  1. alt.atheism,
  2. comp.graphics,
  3. comp.os.ms-windows.misc,
  4. comp.sys.ibm.pc.hardware,
  5. comp.sys.mac.hardware,
  6. comp.windows.x,
  7. misc.forsale,
  8. rec.autos,
  9. rec.motorcycles,
  10. rec.sport.baseball,
  11. rec.sport.hockey,

---

[5]Let $\kappa = (1,1,2,2)^{\mathrm{T}}$, $\lambda^{(1)} = (1,1,2,2)^{\mathrm{T}}$, and $\lambda^{(2)} = (1,2,3,4)^{\mathrm{T}}$. $\lambda^{(2)}$ is an over-refinement of correct clustering $\lambda^{(1)}$. The mutual information between κ and $\lambda^{(1)}$ is 2 and the mutual information between κ and $\lambda^{(2)}$ is also 2. Our [0,1]-normalized mutual information measure $\phi^{(\mathrm{NMI})}$ penalizes the useless refinement: $\phi^{(\mathrm{NMI})}(\lambda^{(2)}, \kappa) = 2/3$ which is less than $\phi^{(\mathrm{NMI})}(\lambda^{(1)}, \kappa) = 1$.

[6]The data can be found at http://www.at.mit.edu/~jrennie/20Newsgroups/.

12. `sci.crypt`,
13. `sci.med`,
14. `sci.electronics`,
15. `sci.space`,
16. `soc.religion.christian`,
17. `talk.politics.guns`,
18. `talk.politics.mideast`,
19. `talk.politics.misc`,
20. `talk.religion.misc`.

- WKB. From the CMU Web KB Project [116], web pages from the following 10 industry sectors according to Yahoo! were selected: `airline`, `computer hardware`, `electronic instruments and controls`, `forestry and wood products`, `gold and silver`, `mobile homes and rvs`, `oil well services and equipment`, `railroad`, `software and programming`, `trucking`. Each industry contributes about 10% of the pages.
- REU. The Reuters-21578, Distribution 1.0.[7] We use the primary topic keyword as the category. There are 82 unique primary topics in the data. The categories are highly imbalanced.

The data sets encompass several text styles. For example, WKB documents vary significantly in length: some are in the wrong category, some are dead links or have little content (e.g., are mostly images). Also, the hub pages that Yahoo! refers to are usually top-level branch pages. These tend to have more similar bag-of-words content across different classes (e.g., contact information, search windows, welcome messages) than news content-oriented pages. In contrast, the content of REU is well-written news agency messages. However, they often belong to more than one category.

Words were stemmed using Porter's suffix stripping algorithm [170] in YAH and REU. For all data sets, words occurring on average between 0.01 and 0.1 times per document were counted to yield the term-document matrix. This excludes stop words such as `a`, and very generic words such as `new`, as well as too rare words such as `haruspex`.

6.2 Summary of Results

In this section, we present and compare the results of the 11 approaches on the four document data sets. Clustering quality is understood in terms of mutual information and balance. For each data set we set the number of clusters k to be twice the number of categories g, except for the REU data set where we used $k = 40$ since there are many small categories. Using a greater number of clusters than classes allows multimodal distributions for each class. For example, in an XOR like problem, there are two classes, but four clusters.

---

[7]Available from Lewis at www.research.att.com/~lewis.

Let us first look at a representative result to illustrate the behavior of some algorithms and our evaluation methodology. In Fig. 3, confusion matrices illustrating quality differences of RND, KM E, KM C, and GP C approaches on a sample of 800 documents from N20 are shown. The horizontal and the vertical axes correspond to the categories and the clusters, respectively. Clusters are sorted in increasing order of dominant category. Entries indicate the number $n_\ell^{(h)}$ of documents in cluster ℓ and category h by darkness. Expectedly, random partitioning RND results in indiscriminating clusters with a mutual information score $\phi^{(\mathrm{NMI})} = 0.16$. The purity score $\phi^{(\mathrm{P})} = 0.16$ indicates that on average the dominant category contributes 16% of the objects in a cluster. However, since labels are drawn from a uniform distribution, cluster sizes are somewhat balanced with $\phi^{(\mathrm{BAL})} = 0.63$. KM E delivers one large cluster (cluster 15) and many small clusters with $\phi^{(\mathrm{BAL})} = 0.03$. This strongly imbalanced clustering is characteristic of KM E on high-dimensional sparse data and is problematic because it usually defeats certain application specific purposes such as browsing. It also results in subrandom quality $\phi^{(\mathrm{NMI})} = 0.11$ ($\phi^{(\mathrm{P})} = 0.17$). KM C results are good. A "diagonal" can be clearly seen in the confusion matrix. This indicates that the clusters align with the ground truth categorization, which is reflected by an overall mutual information $\phi^{(\mathrm{NMI})} = 0.35$ ($\phi^{(\mathrm{P})} = 0.38$). Balancing is good as well with $\phi^{(\mathrm{BAL})} = 0.45$. GP C exceeds KM C in both aspects with $\phi^{(\mathrm{NMI})} = 0.47$ ($\phi^{(\mathrm{P})} = 0.48$) as well as balance $\phi^{(\mathrm{BAL})} = 0.95$. The "diagonal" is stronger and clusters are very balanced.

The rest of the results are given in a summarized form instead of the more detailed treatment in the example mentioned earlier, since the comparative

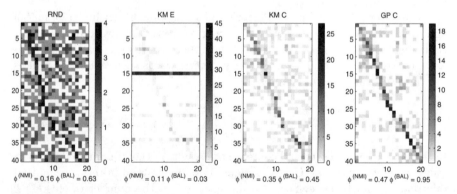

Fig. 3. Confusion matrices illustrating quality differences of RND, KM E, KM C, and GP C approaches on a sample of 800 documents from N20. Matrix entries indicate the number $n_\ell^{(h)}$ of documents in cluster ℓ (row) and category h (column) by darkness. Clusters are sorted in ascending order of their dominant category. KM E delivers one large cluster and shows subrandom quality $\phi^{(\mathrm{NMI})}$. KM C results are good, but are exceeded by GP C in terms of mutual information $\phi^{(\mathrm{NMI})}$ as well as balance $\phi^{(\mathrm{BAL})}$

trends are very clear even at this macrolevel. Some examples of detailed confusion matrices and pairwise t-tests can be found in our earlier work [413].

For a systematic comparison, ten experiments were performed for each of the random samples of sizes 50, 100, 200, 400, and 800. Figure 4 shows performance curves in terms of (relative) mutual information comparing ten algorithms on four data sets. Each curve shows the *difference* $\Delta\phi^{(\mathrm{NMI})}$ in mutual information-based quality $\phi^{(\mathrm{NMI})}$ compared to random partitioning for five sample sizes (at 50, 100, 200, 400, and 800). Error bars indicate ± 1 standard deviations over ten experiments. Figure 5 shows quality in terms of balance for four data sets in combination with ten algorithms. Each curve shows the cluster balance $\phi^{(\mathrm{BAL})}$ for five sample sizes (again at 50, 100, 200, 400, and 800). Error bars indicate ± 1 standard deviations over ten experiments. Figure 6 summarizes the results on all four data sets at the highest sample size level ($n = 800$). We also conducted pairwise t-tests at $n = 800$ to ensure differences in average performance are significant. For illustration and brevity, we chose to show mean performance with standard variation bars rather than the t-test results (see our previous work [413]).

First, we look at quality in terms of mutual information (Figs. 4 and 6a). With increasing sample size n, the quality of clusterings tends to improve. Nonmetric (cosine, correlation, Jaccard) GP approaches work best on text data followed by nonmetric KM approaches. Clearly, a nonmetric, e.g., dotproduct based similarity measure is necessary for good quality. Due to the conservative normalization, depending on the given data set the maximum obtainable mutual information (for a perfect *classifier*!) tends to be around 0.8–0.9. A mutual information-based quality around 0.4 and 0.5 (which is approximately 0.3–0.4 better than random at $n = 800$) is an excellent result.[8] HP constitutes the third tier. Euclidean techniques including SOM perform rather poorly. Surprisingly, the SOM still delivers significantly better than random results despite the limited expressiveness of the implicitly used Euclidean distances. The success of SOM is explained by the fact that the Euclidean distance becomes locally meaningful once the cell centroids are locked onto a good cluster.

All approaches behaved consistently over the four data sets with only slightly different scale caused by the different data sets' complexities. The performance was best on YAH and WKB followed by N20 and REU. Interestingly, the gap between GP and KM techniques is wider on YAH than on WKB. The low performance on REU is probably due to the high number of classes (82) and their widely varying sizes.

In order to assess those approaches that are more suitable for a particular amount of objects n, we also looked for intersects in the performance curves

---

[8] For verification purposes we also computed entropy values for our experiments and compared with, e.g., [463] to ensure validity.

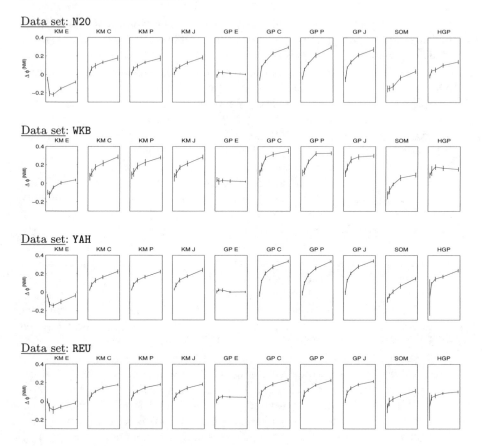

Fig. 4. Mutual information performance curves comparing ten algorithms on four data sets. Each curve shows the difference in mutual information-based quality $\phi^{(\mathrm{NMI})}$ compared to random for five sample sizes (at 50, 100, 200, 400, and 800). Error bars indicate ± 1 standard deviations over ten experiments

of the top algorithms (nonmetric GP and KM, HGP).[9] In our experiments, the curves do *not* intersect indicating that ranking of the top performers does not change in the range of dataset sizes considered.

In terms of balance (Figs. 5 and 6b) the advantages of GP are clear. GP explicitly tries to achieve balanced clusters ($n = 800 : \phi^{(\mathrm{BAL})} \approx 0.9$). The second tier is HGP, which is also a balanced technique ($n = 800 : \phi^{(\mathrm{BAL})} \approx 0.7$) followed by nonmetric KM approaches ($n = 800 : \phi^{(\mathrm{BAL})} \approx 0.5$). Poor balancing

---

[9]Intersections of performance curves in classification (learning curves) have been studied recently, e.g., in [359].

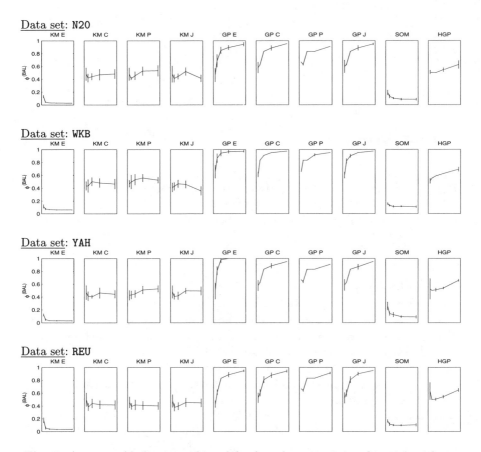

Fig. 5. Amount of balancing achieved for four data sets in combination with ten algorithms. Each curve shows the cluster balance $\phi^{(\mathrm{BAL})}$ for five sample sizes (at 50, 100, 200, 400, and 800). Error bars indicate ± 1 standard deviations over ten experiments.

is shown by SOM and Euclidean KM ($n = 800 : \phi^{(\mathrm{BAL})} \approx 0.1$). Interestingly, balancedness does not change significantly for the KM-based approaches as the number of samples n increases. GP-based approaches quickly approach perfect balancing as would be expected since they are explicitly designed to do so.

Nonmetric GP is significantly better in terms of mutual information as well as balance. There is no significant difference in performance amongst the nonmetric similarity measures using cosine, correlation, and extended Jaccard. Euclidean distance-based approaches do not perform better than random clustering.

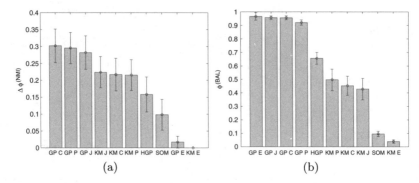

Fig. 6. Comparison of cluster quality in terms of **(a)** mutual information and **(b)** balance on average over four data sets with ten trials each at 800 samples. Error bars indicate ±1 standard deviation. Graph partitioning is significantly better in terms of mutual information as well as in balance. Euclidean distance-based approaches do not perform better than random clustering

7 Conclusions

This work provides a mutual information-based comparison of several similarity-based clustering approaches to clustering of unannotated text *across* several similarity measures. It also provides a conceptual assessment of a variety of similarity measures and evaluation criteria.

The comparative results indicate that for the similarity measures considered, graph partitioning is better suited for word frequency-based clustering of web documents than generalized KM, HGP, and SOM. The search procedure implicit in GP is far less local than the hill-climbing approach of KM. Moreover, it also provides a way to obtain clusters of comparable sizes and exhibit a lower variance in results. Note that while this extra constraint is helpful for datasets that are reasonably balanced, it can degrade results when the classes are highly skewed. With regard to the appropriateness of various distance/similarity measures, it was very clear that metric distances such as the L_2 norm (Euclidean distance) are not appropriate for the high-dimensional, sparse domains that characterize text documents. Cosine, correlation, and extended Jaccard measures are much more successful and perform comparably in capturing the similarities implicitly indicated by manual categorizations of document collections. Note that all three measures tune to different degrees to the directional properties of the data, which is the likely reason for their effectiveness. This intuition is supported by the recent development of a generative model using mixture of von Mises–Fisher distributions from directional statistics and tailored for high-dimensional data, which has been applied to text clustering with clearly superior results [43]. Such generative models are also attractive since their computational complexity can be linear in the number of objects, as compared a mimimum of quadratic complexity for any similarity-based method that involves a comparison between each pair of objects.

Since document clustering is currently a popular topic, a comparative study such as that undertaken in this chapter is by nature an unfinished one as new techniques and aspects emerge regularly. For example, a recent paper introduces a similarity measure based on the number of neighbors two points share, and shows promising results on earth sciences data and word clustering [149]. It will be interesting to see how suitable this measure is for clustering a variety of text collections.

Acknowledgments

We thank Raymond Mooney and Inderjit Dhillon for helpful comments. This work was supported in part by NSF under grants IIS-0307792 and IIS-0325116.

Clustering Very Large Data Sets with Principal Direction Divisive Partitioning

D. Littau and D. Boley

Summary. We present a method to cluster data sets too large to fit in memory, based on a Low-Memory Factored Representation (LMFR). The LMFR represents the original data in a factored form with much less memory, while preserving the individuality of each of the original samples. The scalable clustering algorithm Principal Direction Divisive Partitioning (PDDP) can use the factored form in a natural way to obtain a clustering of the original dataset.

The resulting algorithm is the PieceMeal PDDP (PMPDDP) method. The scalability of PMPDDP is demonstrated with a complexity analysis and experimental results. A discussion on the practical use of this method by a casual user is provided.

1 Introduction

One of the challenges in data mining is the clustering of very large data sets. We define a very large data set as a data set that will not fit into memory at once. Many clustering algorithms require that the data set be scanned many times during the clustering process. If the data cannot fit into memory, then the data must be repeatedly rescanned from disk, which can be expensive.

One approach to clustering large data sets is to adapt clustering algorithms suitable for small data sets to much larger data sets. There are two popular methods used to adapt clustering algorithms to large data sets. The first technique is to extract a subsample of the data, such that the subsample is small enough to fit into available memory and be clustered. Other techniques to accelerate the clustering process are often applied at the same time. Once a clustering is obtained, the remaining data points can be assigned to the clusters with the closest centroid. The major drawbacks to sampling are that it can be difficult to know if a given subsample is a representative sample and therefore provides an accurate clustering, and that the outliers will usually be ignored.

The second technique commonly used to adapt clustering algorithms to large data sets, as in [85, 121, 459], is to approximate a given data item by assigning it to a single representative vector. One representative vector may take the place of an arbitrary number of data items. Once a data item has been assigned to a representative, it is no longer possible to differentiate it from any other data item assigned to the same vector. Thus, the resolution of any clustering of the data is limited by the granularity of the representatives.

We propose an alternate approach to adapt the principal direction divisive partitioning (PDDP) clustering method [76] to very large data sets. We create a low-memory factored representation (LMFR) of the data, and then cluster the LMFR using PDDP. Every data point has a unique representation in the LMFR, and every data point is examined during the construction of the LMFR. The LMFR is constructed piecewise using samples of the data, such that the samples will fit into memory. The samples are selected without replacement, and selection continues until the data set is exhausted. Once an approximation has been constructed for each sample, the approximations are assembled into an LMFR representing the entire data set.

The LMFR avoids what we claim are the major drawbacks to other techniques. All data are examined during the construction of the LMFR, which is not the case when subsamples are clustered as a representative of the entire data set. Each data item has a unique representation in the LMFR, so the granularity of the clustering can be finer than that achieved by a method which assigns many data items to a single representative vector. Every data item is examined and participates in the construction of the LMFR, so outliers will not be ignored. Furthermore, since the method is deterministic, we need not be concerned that other, perhaps better clusterings could be constructed.

The remainder of the chapter is as follows. First, we provide some background on a few of the methods available to cluster large data sets. Next, we describe the technique used to construct the LMFR. Then, we describe how the original representation of the data can be easily replaced by the LMFR in the PDDP method, a process we call Piecemeal PDDP (PMPDDP). Finally we show some experimental results, which demonstrate that the clustering quality of the PMPDDP method is similar to PDDP, and that PMPDDP maintains the scalability of PDDP.

2 Background

The problem of clustering very large data sets is an active area of research. Many approaches adapt existing clustering methods such as hierarchical agglomeration [202] and k-means [143, p. 201] to much larger sets. There are also clustering methods that were specifically designed from the ground up to be used for large data sets. Note that the following is a sketch of some clustering methods for large data sets and is not intended to be taken as exhaustive.

2.1 Sampling

Before we describe any specific methods, we describe sampling. Sampling is a general approach to extending a clustering method to very large data sets. A sample of the data is selected and clustered, which results in a set of cluster centroids. Then, all data points are assigned to the closest centroid. Many large data set clustering methods use sampling to overcome time and memory limitations.

2.2 Hierarchical Agglomeration and Its Variants

Hierarchical agglomeration [202] produces a hierarchy of clusters, such that any given level of cluster refinement can be selected from the results. It starts with singleton clusters and produces the hierarchy of clusters by successively merging the two clusters that are closest. Typically, the distances between clusters are determined by computing the distance from every point in a given cluster to every other point in every other cluster. The expense of computing the distances between all points is the most serious drawback to the method.

Scatter/Gather [121] speeds up agglomeration by dividing the data into buckets and agglomerating individual buckets until the number of clusters in a given bucket is reduced by a specific amount. The clusters are replaced by their weighted centroids, and the centroids from all buckets are then placed in a smaller set of buckets and agglomerated again. The process continues until a specified number of centroids are created, after which all data are assigned to the closest centroid. While this method was specified as a speed-up for data sets that would fit into memory, combining Scatter/Gather with sampling would make it appropriate for large data sets. Alternately, the amount of data in a given bucket could be sized to fit into memory, and only one bucket of data would appear in memory and be agglomerated at a given time. The resulting centroids could be saved to disk, and another bucket of data could then be loaded and agglomerated, and so on. This would require some additional disk access, but would result in a method that could work for arbitrarily large data sets.

CURE [209] adapts hierarchical agglomeration by using a small set of well-scattered points to compute the distances between clusters, rather than considering all the points in a cluster when computing the distances between clusters. This significantly speeds up the procedure. Subsampling the data was also specified when the data set was too large.

There are other extensions of hierarchical agglomeration. For instance, [174] uses maximum likelihood determined by a multivariate Gaussian model to decide the two clusters that should be merged. The work in [458] uses a heap to store the distances between all pairs to speed up access to distance information. Refinement of the clusters to increase quality is described in [261]. While these methods were designed to enhance the speed and quality of hierarchical agglomeration, combining them with subsampling the data would make them suitable for clustering very large data sets.

2.3 k-Means and Its Variants

k-Means produces clusters using an iterative method. A random set of starting centroids is selected from the data set, and all data points are assigned to the closest center. Then, new centroids are computed using the data points in each cluster, and again all data points are assigned to the closest centroid. The process continues until there is no further data movement. Multiple passes with random restarts are usually performed to ensure a good clustering has been found.

One adaptation of k-means to very large data sets is provided in [85]. Samples are drawn from the data set, without replacement, and clustered. The data points in a given cluster are replaced by a representative, which is much like a weighted centroid, but provides a bit more information. This is done for all current clusters. Then more samples are drawn from the data set and are clustered along with the weighted centroids. The process continues until the data set is exhausted or the centroids stop moving.

It is difficult to know good choices for starting centers for k-means. Instead of repeating k-means with random restarts, [84] provides a technique to generate good candidate centers to initialize k-means. The method works by selecting some random samples of the data and clustering each random sample separately using k-means. The centroids from each clustering are then gathered into one group and clustered to create a set of initial centers for a k-means clustering of the entire data set.

There are other variants of k-means. The work in [357] uses a k–d tree to organize summaries of the data. It is fast but does not perform well for dimensions higher than eight. The work in [20] used a k–d tree to cut down on the number of distance computations required, though it is not clear if the application is limited to spatial data. We assume it is, since they used the same kind of data structure as in [357], and their experiments were conducted on low-dimension data. Therefore these methods are more appropriate for large low-dimension data sets.

Very fast k-means (VFKM) [141] takes a different approach from other clustering methods. The stated desire is to produce a model (clustering) using a finite amount of data that cannot be distinguished from a model constructed using infinite data. The error is bounded by comparing the centroids resulting from different k-means clusterings using different sample sizes. Stated in a very simplified manner, if the centroids from the different clusterings are within a specified distance of each other, they are considered to be the correct centroids. Otherwise a new, larger sample is drawn from the data set and clustered, and the resulting centroids are compared to the those obtained from the previous run. The authors suggest that this method is not a reasonable approach unless the database being clustered contains millions of items.

2.4 Summary of Cited Clustering Methods

Most of the extensions of hierarchical agglomeration were designed to speed up the process for data sets that can fit into memory. Sampling was indicated when the data sets grew too large. Sampling the data ignores outliers, which may be interesting data items in some circumstances. Also, it is difficult to know whether a truly representative sample has been drawn from the data set.

The extensions of k-means to large data sets either drew samples or assigned many data points to one representative vector. Using one vector to approximate many data items, as in [85, 121, 459], is a relatively popular technique when constructing approximations to the data. However, once the assignments have been made, there is no way to distinguish between the data items assigned to a given representative. The resolution of any clustering of the data is limited by the resolution of the representatives.

In the clustering method we present in this chapter, no sampling of the data is necessary. All data items are exposed to the method. Each data item has a unique representation in the approximation we construct. Therefore, the resolution of the clustering is not limited by the approximation of the data. We believe these differences result in a method which provides a useful alternative to other large data set clustering methods.

3 Constructing a Low-Memory Factored Representation

The LMFR is comprised of two matrices. The first matrix contains representative vectors, and the second contains data loadings. The representative vectors are the centroids obtained from a clustering of the data. The data loadings are a least-squares approximation to each data item using a small number of selected representative vectors.

Since the data set is assumed to be too large to fit into memory, we divide it up into smaller samples called *sections*. Each data item from the original representation appears once and only once across all sections. We individually compute an LMFR for each section.

First, we describe the method used to obtain the LMFR for one section of data. Then, we describe how we assemble the LMFRs for each section into one LMFR, which represents the entire data set.

3.1 Constructing an LMFR for One Section

Suppose we have an $n \times m$ matrix $\mathbf{A}$ of data samples, such that $\mathbf{A}$ comfortably fits into memory at once. We compute the factored representation

$$\mathbf{A} \approx \mathbf{C}_A \mathbf{Z}_A, \tag{1}$$

where $\mathbf{C}_A$ is an $n \times k_c$ matrix of representative vectors and $\mathbf{Z}_A$ is a $k_c \times m$ matrix of data loadings. Each column $\mathbf{z}_i$ of $\mathbf{Z}_A$ approximates the corresponding column $\mathbf{a}_i$ of $\mathbf{A}$ using a linear combination of the vectors in $\mathbf{C}_A$.

The first step in computing this factored form of $\mathbf{A}$ is to obtain the matrix of representative vectors $\mathbf{C}_A$. To accomplish this, we partition $\mathbf{A}$ into k_c clusters and compute the k_c centroids of the clusters. These centroids are collected into an $n \times k_c$ matrix $\mathbf{C}_A$,

$$\mathbf{C}_A = [\mathbf{c}_1 \ \mathbf{c}_2 \ \cdots \ \mathbf{c}_{k_c}]. \tag{2}$$

We use the PDDP method to compute the clustering of $\mathbf{A}$ and therefore obtain $\mathbf{C}_A$, since PDDP is fast and scalable. In principle, any clustering method could be used to obtain the components of $\mathbf{C}_A$.

The matrix of data loadings $\mathbf{Z}_A$ is computed one column at a time. In approximating each column $\mathbf{a}_i$ of $\mathbf{A}$, we use only a small number (k_z) of the representatives in $\mathbf{C}_A$. Therefore, each column $\mathbf{z}_i$ in $\mathbf{Z}_A$ has only k_z nonzero entries. For example, to approximate $\mathbf{a}_i$, we choose the k_z columns in $\mathbf{C}_A$ which are closest in Euclidean distance to $\mathbf{a}_i$ and implicitly collect them into an $n \times k_z$ matrix $\mathbf{C}_i$. Then the nonzero entries in the column $\mathbf{z}_i$ are obtained by solving for the k_z-vector $\hat{\mathbf{z}}_i$:

$$\hat{\mathbf{z}}_i = \arg \min_{\mathbf{z}} ||\mathbf{a}_i - \mathbf{C}_i \mathbf{z}||_2^2. \tag{3}$$

If the k_z vectors in $\mathbf{C}_i$ are linearly independent, we use the normal equations with the Cholesky decomposition to solve the least-squares problem. If the normal equations fail, we use the more expensive SVD to get the least-squares approximation of the data item. Even though there has been no attempt to create orthogonal representative vectors, in the majority of cases the normal equations are sufficient to solve the least-squares problem. The LMFR algorithm is shown in Fig. 1.

When $k_z = k_c$, this factorization of $\mathbf{A}$ is essentially identical to the *concept decomposition* [136], except that we use PDDP to obtain the clustering rather than spherical k-means. We typically select a value for k_z such that $k_z \ll k_c$, which can result in significant memory savings. Since the memory savings are dependent on $\mathbf{Z}$ being sparse, we also require the condition that $k_z \ll n$. Thus a low-dimension matrix is not a good candidate for this factorization technique from a memory-savings standpoint.

To obtain memory savings, it is also necessary to control the size of $\mathbf{C}_A$, which is done by making k_c as small as possible. There is a trade-off between

---

Algorithm LMFR.

0. **Start** with a $n \times m$ matrix $\mathbf{A}$, where each column of $\mathbf{A}$ is a data item, and set the values for k_c, the number of representative vectors in $\mathbf{C}$, and k_z, the number of representatives used to approximate each data item.
1. **Partition** $\mathbf{A}$ into k_c clusters
2. **Assemble** the k_c cluster centroids from step 1 into an $n \times k_c$ matrix $\mathbf{C}_A$ ((2) in the text).
3. **For** $i = 1, 2, \ldots, m$ **do**
4.   **Find** the k_z columns in $\mathbf{C}_A$ closest to $\mathbf{a}_i$
5.   **Collect** the k_z columns found in step 4 as the $n \times k_z$ matrix $\mathbf{C}_i$
6.   **Compute** $\hat{\mathbf{z}}_i = \arg \min_{\mathbf{z}} ||\mathbf{a}_i - \mathbf{C}_i \mathbf{z}||_2^2$
7.   **Set** the ith column of $\mathbf{Z}_A = \hat{\mathbf{z}}_i$
8. **Result** $\mathbf{C}_A$ and $\mathbf{Z}_A$, which represent a factorization of $\mathbf{A}$

Fig. 1. LMFR algorithm

the two parameters k_c and k_z, since for a given amount of memory available to contain the LMFR $\mathbf{C}_A\mathbf{Z}_A$, increasing one of the parameters requires decreasing the other.

3.2 Constructing an LMFR of a Large Data Set

Once an LMFR has been computed for each section, they are assembled into a single factored representation of the entire original data set. A graphical depiction of this technique is shown in Fig. 2. What follows is a formal definition of the entire process of constructing the LMFR of a large data set.

We consider the original representation of the data set as an $n \times m$ matrix $\mathbf{M}$, such that $\mathbf{M}$ will not fit into memory at once. We seek the single factored

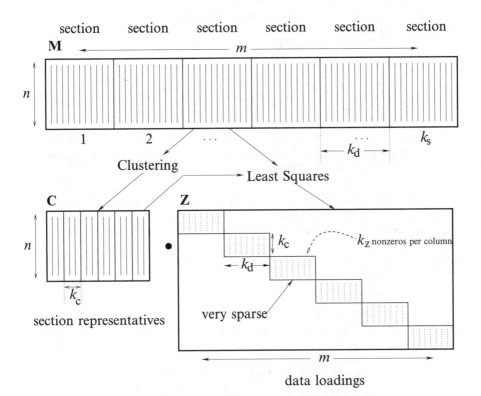

Fig. 2. Construction details of the low-memory representation. $\mathbf{M}$ is divided into k_s sections, and the low-memory representation of each section is computed without referring to any other section. Each section is associated with a subdivision of $\mathbf{C}$ and $\mathbf{Z}$. The columns of a subdivision of $\mathbf{C}$ are the cluster centroids resulting from a clustering of the associated section. A column of a subdivision of $\mathbf{Z}$ is computed with a least-squares approximation to the corresponding column of $\mathbf{M}$, using the k_z closest centroids from the associated subdivision of $\mathbf{C}$

representation $\mathbf{CZ}$ such that

$$\mathbf{M} \approx \mathbf{CZ}, \tag{4}$$

where $\mathbf{C}$ and $\mathbf{Z}$ will fit into memory and can be used to cluster the data in $\mathbf{M}$. Since $\mathbf{M}$ cannot fit into memory at once, $\mathbf{M}$ is divided into k_s disjoint sections,

$$\mathbf{M} = [\mathbf{M}_1 \; \mathbf{M}_2 \; \cdots \; \mathbf{M}_{k_s}], \tag{5}$$

such that each section $\mathbf{M}_j$ of $\mathbf{M}$ will fit into memory. This partitioning of $\mathbf{M}$ is virtual since we assume only one section $\mathbf{M}_j$ will be in memory at any given instance. We also assume that the ordering of the columns of $\mathbf{M}$ is unimportant. We can now construct an LMFR

$$\mathbf{M}_j \approx \mathbf{C}_j \mathbf{Z}_j \tag{6}$$

for each section $\mathbf{M}_j$ of $\mathbf{M}$ using the technique from Sect. 3.1. After computing an approximation (6) for each section of data, they can be assembled into the two-matrix system

$$\mathbf{C} = \begin{bmatrix} \mathbf{C}_1 & \mathbf{C}_2 & \cdots & \mathbf{C}_{k_s} \end{bmatrix},$$
$$\mathbf{Z} = \begin{bmatrix} \mathbf{Z}_1 & & & \\ & \mathbf{Z}_2 & & \mathbf{0} \\ & & \ddots & \\ \mathbf{0} & & & \mathbf{Z}_{k_s} \end{bmatrix}, \tag{7}$$

where $\mathbf{C}$ has the dimension $n \times k_s k_c$ and $\mathbf{Z}$ has the dimension $k_s k_c \times n$. We call this system a general LMFR. The parameters used to construct the general LMFR are summarized in Table 1.

Note that the idea of processing the data in separate pieces and assembling the results has been done previously in the context of principal component analysis [255]. However, in that case the application was intended for situations in which different sets of attributes for a given data point were distributed across separate databases. The LMFR construction method is designed to process data points that have all attributes present.

Table 1. Definition of the parameters used in constructing a general LMFR

Parameter	Description
m	Total number of data items
n	Number of attributes per data item
γ_1	Fill fraction of the attributes in $\mathbf{M}$
k_s	Number of sections
k_d	Number of data items per section
k_c	Number of centroids per section
k_z	Number of centroids approximating each data item

3.3 Applications of the LMFR

The only application of the LMFR which we cover in this work is using the LMFR to extend PDDP to large data sets. However, this is not the only successful application of the LMFR in data mining. In [319] we showed an adaptation of the LMFR to general stream mining applications. The LMFR allows for more of the stream to be exposed to a given stream mining method at once.

Another application we have investigated is using the LMFR for document retrieval [317]. We demonstrated that we could construct an LMFR of a given data set that had better precision vs. recall than an SVD of a specific rank, while taking less time to construct and occupying less memory than the SVD. Specifically, the LMFR with $k_z = 5$ and $k_c = 600$ for a 7,601-item data set took 187 s to construct and occupied 11.32 MB of memory, while a rank 100 SVD took 438 s to construct and occupied 40.12 MB of memory. Given the advantage in construction time and memory used, the LMFR appears to be a viable alternative to the SVD for document retrieval.

4 Complexity of the LMFR

We now provide a complexity analysis of the cost of computing the LMFR. To make the explanation simpler, we make the following assumptions: the data set represented by $\mathbf{M}$ is evenly distributed among the sections so that k_c and k_d are the same for each section, and k_z is the same for each section k_s, $k_d k_s = m$, and $m \gg n$. These are not necessary conditions to construct an LMFR, but they do make the explanation clearer.

4.1 Cost of Obtaining the Section Representatives

The first step in computing the LMFR $\mathbf{C}_j \mathbf{Z}_j$ for a given section of data $\mathbf{M}_j$ is to obtain the section representatives that comprise $\mathbf{C}_j$. These are found via a clustering of $\mathbf{M}_j$. We assume that PDDP will be used to obtain the clustering. To simplify the analysis, we assume that we will create a perfectly balanced binary tree. This means that all the leaf clusters in a given "level" will have the same cardinality, and that all the clusters on a given level will be split before any clusters on the next level.

The major cost of computing the PDDP clustering is that of computing the principal direction of the data in the current cluster being split. The principal direction is determined by the rank 1 SVD of the cluster. The rank 1 SVD is computed using the iterative procedure developed by Lanczos. The major cost in finding the rank 1 SVD is computing a matrix–vector product of the form $\mathbf{M}_j \mathbf{v}$ twice each iteration.

PDDP starts with the root cluster, which is all the data being clustered. In this case, the root cluster is the $n \times k_d$ matrix $\mathbf{M}_j$, where n is the number of attributes and k_d is the number of data items in the section. Computing the product of one row in $\mathbf{M}_j$ with a right vector $\mathbf{v}$ takes k_d multiplications and additions. There are n rows in $\mathbf{M}_j$. Therefore, the cost of computing a single matrix–vector product is

$$\gamma_1 n k_d, \tag{8}$$

where γ_1 is the fill fraction of $\mathbf{M}_j$. If the data in $\mathbf{M}$ are dense, $\gamma_1 = 1$. The overall cost of determining the principal direction of the root cluster $\mathbf{M}_j$ is

$$c_1 \gamma_1 n k_d, \tag{9}$$

where c_1 is a constant encapsulating the number of matrix–vector products computed before convergence.

After splitting the root cluster, we have two leaf clusters. Due to our assumption that we are creating a perfectly balanced binary tree, each of the two current leaf clusters contains the same number of data items, and the next two clusters chosen to be split will be the two current leaf clusters. Therefore, the cost of splitting the next two clusters is

$$2 c_1 \gamma_1 n \left(\frac{k_d}{2} \right) = c_1 \gamma_1 n k_d, \tag{10}$$

which is the same as the cost of computing the splitting of the root cluster. The PDDP tree now contains four leaf clusters. The cost of splitting these four leaf clusters is

$$4 c_1 \gamma_1 n \left(\frac{k_d}{4} \right) = c_1 \gamma_1 n k_d. \tag{11}$$

This result and the previous result indicate that the cost of computing a given level in the binary tree is the same for all levels. Every new level created in the perfectly balanced binary tree increases the number of leaf clusters by a power of 2. This progression is shown in Fig. 3.

The cost of obtaining the k_c section representatives for the section $\mathbf{M}_j$ is

$$c_1 \gamma_1 n k_d \log_2 (k_c), \tag{12}$$

assuming that the number of section representatives k_c is an integer power of 2. If we have a total of k_s sections with k_d data points per section, and we obtain the same number of section representatives k_c for each section, the total cost of obtaining all section representatives for the entire data set will be

$$cost_C = c_1 \gamma_1 n k_s k_d \log_2 (k_c) = c_1 \gamma_1 n m \log_2 (k_c). \tag{13}$$

For clarity, we reproduce all the assumptions involved in the formulation, as well as the final result for the cost of computing the section representatives, in Fig. 4.

Number of clusters	Cost
2	$c_1\gamma_1 n k_d$
4	$c_1\gamma_1 n k_d + 2c_1\gamma_1 n\left(\dfrac{k_d}{2}\right)$
	$= 2c_1\gamma_1 n k_d$
8	$c_1\gamma_1 n k_d + 2c_1\gamma_1 n\left(\dfrac{k_d}{2}\right) + 4c_1\gamma_1 n\left(\dfrac{k_d}{4}\right)$
	$= 3c_1\gamma_1 n k_d$
16	$c_1\gamma_1 n k_d + 2c_1\gamma_1 n\left(\dfrac{k_d}{2}\right) + 4c_1\gamma_1 n\left(\dfrac{k_d}{4}\right) + 8c_1\gamma_1 n\left(\dfrac{k_d}{8}\right)$
	$= 4c_1\gamma_1 n k_d$
k_c	$c_1\gamma_1 n k_d \log_2(k_c)$

Fig. 3. Complexity for the PDDP tree computation, shown for the number of clusters computed for a given section of data. The value of k_c is assumed to be an integer power of 2, and the tree is assumed to be perfectly balanced

> **Assumptions:**
> 1. PDDP is used to obtain the section representatives
> 2. A perfectly balanced binary tree is created
> 3. each section has the same value of k_d and k_c
> 4. $k_d k_s = m$
> 5. k_c is an integer power of two
>
> **Result:** Cost of obtaining $\mathbf{C}$ is
> $$cost_\mathbf{C} = c_1\gamma_1 nm \log_2(k_c)$$

Fig. 4. Summary of the cost of obtaining $\mathbf{C}$

4.2 Computing the Data Loadings

Computing the data loadings in $\mathbf{Z}_j$ is a multistep process. To find the least-squares approximation to a given data item $\mathbf{x}_i$ in $\mathbf{M}_j$, it is necessary to find the distance from $\mathbf{x}_i$ to every section representative $\mathbf{c}_l$ in $\mathbf{C}_j$, select the k_z section representatives $\mathbf{c}_l$ that are closest to $\mathbf{x}_i$, and compute the least-squares approximation using the normal equations.

Computing the distance from $\mathbf{x}_i$ to a single representative in $\mathbf{C}_j$ requires $\gamma_1 n$ multiplications and subtractions. Since there are k_c representatives in $\mathbf{C}_j$, the total cost of computing the distances for one data item $\mathbf{x}_i$ is $\gamma_1 n k_c$. We assume that the number of representatives k_z used to approximate x_i is very small, so that it will be less expensive to directly select the k_z closest representatives, rather than sorting the distances first. Therefore, it takes $k_c k_z$ searches through the representatives to find the k_z closest representatives, which are used to form the $n \times k_z$ matrix $\mathbf{C}_i$ as in Sect. 3.1.

Assumptions:
  1. $k_z \ll k_c$, so direct search for k_z closest representatives
     in $\mathbf{C}$ is less expensive than sorting
  2. same value for k_z is used for all sections
  3. normal equations are used to obtain least-squares
  4. additional lower-order terms from least squares are ignored
Result: Cost of obtaining $\mathbf{Z}$ is
$$cost_Z = m\left(\gamma_1 nk_c + k_c k_z + \gamma_1 nk_z^2 + \tfrac{1}{3}k_z^3\right)$$

Fig. 5. Summary of the cost of obtaining $\mathbf{Z}$

The final step is to compute the least-squares approximation for each data item using the k_z centers obtained in the previous step. The normal equations are used to obtain the least-squares approximation. The cost of computing a least-squares approximation for the $n \times k_z$ system is:

$$\gamma_1 nk_z^2 + \frac{1}{3}k_z^3,$$

if we ignore lower-order terms. The total combined cost for obtaining the loadings for one data item $\mathbf{x}_i$ is

$$\gamma_1 nk_c + k_c k_z + \gamma_1 nk_z^2 + \frac{1}{3}k_z^3, \tag{14}$$

and the cost of obtaining all data loadings for all sections is

$$cost_Z = m\left(\gamma_1 nk_c + k_c k_z + \gamma_1 nk_z^2 + \frac{1}{3}k_z^3\right). \tag{15}$$

The assumptions and final cost for computing the data loadings, which comprise the $\mathbf{Z}$ matrix, are shown in Fig. 5.

5 Clustering Large Data Sets Using the LMFR

Now that we have an LMFR of the entire data set, we can replace the original representation of the data with the LMFR to obtain a clustering using PDDP. We call the extension of PDDP to large data sets piecemeal PDDP (PMPDDP). The piecemeal part of the name is from the fact that the LMFR is constructed in a piecemeal fashion, one section at a time, and from the fact that PDDP is used to compute the intermediate clusterings used in the construction of the LMFR.

The PMPDDP clustering method is straightforward. The process is to first construct the LMFR of the data, and then cluster the LMFR using PDDP.

Algorithm PMPDDP.

0. **Start** with a $n \times m$ matrix $\mathbf{M}$, where each column of $\mathbf{M}$ is a data item, and set the values for k_s, k_c, k_z (see Table 1) and k_f, the number of final clusters computed

1. **Partition** $\mathbf{M}$ into k_s disjoint sections, $|\mathbf{M}_1 \mathbf{M}_2, \ldots, \mathbf{M}_{k_s}|$.

2. **For** $j = 1, 2, \ldots, k_s$ **do**

3.   **Compute** the LMFR (cf. Fig. 1) for the section $\mathbf{M}_j$ using PDDP to compute $\mathbf{C}_j$ (cf. S3.2)

4.. **Assemble** the matrices $\mathbf{C}$ and $\mathbf{Z}$ as in (7) in the text, using all the matrices $\mathbf{C}_j$ and $\mathbf{Z}_j$ from all passes through steps 2-3

5.. **Compute** the PDDP tree with k_f clusters for the entire system $\mathbf{CZ}$.

6. **Result:** A binary tree with k_f leaf nodes forming a partitioning of the entire data set.

Fig. 6. PMPDDP algorithm

The PMPDDP algorithm is shown in Fig. 6. An earlier version of PMPDDP appeared in [318].

PDDP is useful for producing the section representatives in $\mathbf{C}$ because it is fast and scalable. Since we are only interested in finding suitable representatives, we do not require the optimal clustering of the data in a section, just an inexpensive one. More expensive clustering algorithms will probably not significantly alter the results, though of course the values in the factorization would change. However, we could replace PDDP with any other suitable clustering algorithm without difficulty, since when we compute the section representatives we are dealing with a piece of the original data that will fit into memory. k-Means, especially bisecting k-means [407], for example, would be candidate methods to replace PDDP at this stage.

However, when clustering the factored form, the situation is different. Any clustering algorithm that uses a similarity measure, such as the aforementioned k-means method, would require that the data be reconstructed each time a similarity measure was needed. Reconstructing the entire data set at once requires at least as much memory as the original data set, defeating the purpose of the LMFR. Reconstructed sparse data will take up more space than the original data, since the section representatives will be denser than the original data items. The LMFR only saves memory as long as it remains in at factored form. Naturally, small blocks of the original data could be reconstructed on the fly every time a similarity measure is required, but that could add considerable additional expense.

PDDP does not use a similarity measure when determining the clustering. Instead, PDDP uses the principal direction of the data in a cluster to determine how to split that cluster. The principal direction is computed using an iterative procedure developed by Lanczos which computes products of the form:

$$\left(\mathbf{M} - \mathbf{w}\mathbf{e}^{\mathrm{T}}\right)\mathbf{v} = \mathbf{M}\mathbf{v} - \mathbf{w}\left(\mathbf{e}^{\mathrm{T}}\mathbf{v}\right), \quad \text{where } \mathbf{w} = \frac{1}{m}\left(\mathbf{M}\mathbf{e}^{\mathrm{T}}\right), \qquad (16)$$

where $\mathbf{v}$ is some vector. We can replace $\mathbf{M}$ by the factored form $\mathbf{CZ}$, group the products accordingly, and compute:

$$\mathbf{C}\left(\mathbf{Z}\mathbf{v}\right) - \hat{\mathbf{w}}\left(\mathbf{e}^{\mathsf{T}}\mathbf{v}\right), \text{ where } \hat{\mathbf{w}} = \frac{1}{m}\mathbf{C}\left(\mathbf{Z}\mathbf{e}^{\mathsf{T}}\right), \tag{17}$$

and in doing so we never explicitly reconstruct the data. Therefore, the LMFR is well suited to being clustered using the PDDP method.

5.1 Scatter Computation

There is one other aspect of PMPDDP to consider. PDDP normally chooses the next cluster to split based on the scatter values of the leaf clusters. Computing the scatter when clustering the LMFR $\mathbf{CZ}$ would require that the data be reconstructed. For a scatter computation, this could be done in a block-wise fashion without too much difficulty. However, we wish to have a method that does not require reconstruction of the data.

Instead of reconstructing the data to directly compute the scatter, we estimate the scatter. If we could compute all the singular values σ_i, we could compute the exact scatter s as

$$s = \sigma_1^2 + \sigma_2^2 + \cdots + \sigma_n^2. \tag{18}$$

This formula can be rewritten as

$$s = \sigma_1^2 \left(1 + \frac{\sigma_2^2}{\sigma_1^2} + \frac{\sigma_3^2}{\sigma_1^2} + \cdots + \frac{\sigma_n^2}{\sigma_1^2}\right). \tag{19}$$

Now, we can use the two leading singular values to estimate the scatter as

$$s \approx \sigma_1^2 \left(1 + \frac{\sigma_2^2}{\sigma_1^2} + \left(\frac{\sigma_2^2}{\sigma_1^2}\right)^2 + \cdots + \left(\frac{\sigma_2^2}{\sigma_1^2}\right)^{n-1}\right) = \sigma_1^2 \left(\frac{1 - \left(\frac{\sigma_2^2}{\sigma_1^2}\right)^n}{1 - \frac{\sigma_2^2}{\sigma_1^2}}\right), \tag{20}$$

where σ_1 is the leading singular value of the cluster and σ_2 is the next singular value. The estimate assumes that the singular values decrease geometrically, which from empirical observation seems to be a reasonable assumption. Note that if we computed all the singular values, we could compute the exact scatter. However, computing all or even a significant number of the singular values would be prohibitively expensive. A high degree of accuracy is not necessary, since this scatter computation is only used to determine the cluster to be split next. The estimate needs only to be consistent with the data being clustered. Presumably, if the estimate is either too low or too high, the same type of estimation error will exist for all clusters.

The leading singular value is associated with the principal direction, and an estimate of the second singular value is available without much additional cost. Estimating the scatter requires that the principal direction of all leaf clusters needs to be computed, whether they are split or not. We could choose another splitting criterion, but from previous results with PDDP on various data sets, scatter seems to be a very good way to select the clusters to split.

6 Complexity of a PMPDDP Clustering

In the following we develop some formulas for the cost of a general PMPDDP clustering. We use the same assumptions as we did for the analysis used to get the section representatives comprising C (cf. $S4.1$). We assume we will produce a completely balanced binary tree with a number of leaves being an integer power of 2, and that the cost of clustering is basically the cost of obtaining the principal directions, which determine how each cluster is split.

Replacing the original matrix M with the approximation CZ in the PDDP method changes the calculation in the splitting process from a matrix–vector product to a matrix–matrix–vector product. This product can be written as $C(Zv)$, where v is an "generic" $m \times 1$ vector, C is an $n \times k_s k_c$ matrix, and Z is a $k_s k_c \times m$ matrix. Note that we group the product such that the matrix-vector product is computed before multiplying by the other matrix. We must avoid explicitly forming the product CZ, since the result of that product will not fit into memory. Z is a sparse matrix with k_z nonzeroes per column, and therefore the only computation cost with respect to Z is incurred when computing the product of the nonzero elements in Z with the elements in v. We show all the parameters involved in a PMPDDP clustering in Table 2.

Again, we start the analysis with the root cluster. The cost of computing the principal direction of the root cluster is

$$c_2(\gamma_2 nk_s k_c + mk_z), \qquad (21)$$

where γ_2 is the fill fraction of C and c_2 is a constant encapsulating the number of matrix–matrix–vector products required to convergence. The mk_z portion of the formula is the contribution from forming the product Zv, where v is some vector and $nk_s k_c$ is the cost of multiplying C by the resultant of the product Zv.

At this point, we have two leaf clusters. One might be tempted to assume that the expense would follow the same relationship as in regular PDDP, and

Table 2. Definition of the parameters used in PMPDDP

Parameter	Description
m	Total number of data items
n	Number of attributes per data item
γ_1	Fill fraction of the attributes in M
γ_2	Fill fraction of the attributes in C
k_s	Number of sections
k_d	Number of data items per section
k_c	Number of centroids per section
k_z	Number of centroids approximating each data item
k_f	Number of final clusters

that the cost of splitting these two clusters is the same as that of the root cluster, but that is incorrect. The reason for the difference is that while the cost of forming the product $\mathbf{Zv}$ decreases with decreasing cluster size, the cost of multiplying $\mathbf{C}$ by the resultant of the product $\mathbf{Zv}$ does not decrease with decreasing cluster size. As a result, the cost of splitting these two clusters is

$$2c_2\gamma_2 nk_s k_c + 2c_2 \left(\frac{m}{2}\right) k_z = 2c_2\gamma_2 nk_s k_c + c_2 mk_z. \tag{22}$$

It might be possible to reduce the computational expense associated with $\mathbf{C}$ by only considering the columns of $\mathbf{C}$ which actually participate in the product when splitting the leaf clusters. However, there does not appear to be an inexpensive way to determine the columns in $\mathbf{C}$ that would be required at each step, so we leave the method as stated and accept the expense.

Following the pattern to its conclusion, as shown in Fig. 7, we have the result for the cost of clustering $\mathbf{CZ}$,

$$c_2\gamma_2 nk_s k_c(k_f - 1) + c_2 mk_z \log_2(k_f). \tag{23}$$

We have not yet considered the fact that PMPDDP uses the estimated scatter to determine the cluster that is split next. To obtain the estimated

Number of clusters	Cost
2	$c_2\gamma_2 nk_s k_c + c_2 mk_z$
4	$c_2\gamma_2 nk_s k_c + c_2 mk_z + 2c_2\gamma_2 nk_s k_c + 2c_2 \left(\dfrac{m}{2}\right) k_z$
	$= 3c_2\gamma_2 nk_s k_c + 2c_2 k_z m$
8	$c_2\gamma_2 nk_s k_c + c_2 mk_z + 2c_2\gamma_2 nk_s k_c + 2c_2 \left(\dfrac{m}{2}\right) k_z$
	$+ \;\; 4c_2\gamma_2 nk_s k_c + 4c_2 \left(\dfrac{m}{4}\right) k_z$
	$= 7c_2\gamma_2 nk_s k_c + 3c_2 mk_z$
16	$c_2\gamma_2 nk_s k_c + c_2 mk_z + 2c_2\gamma_2 nk_s k_c + 2c_2 \left(\dfrac{m}{2}\right) k_z$
	$+ \;\; 4c_2\gamma_2 nk_s k_c + 4c_2 \left(\dfrac{m}{4}\right) k_z$
	$+ \;\; 8c_2\gamma_2 nk_s k_c + 8c_2 \left(\dfrac{m}{8}\right) k_z$
	$= 15c_2\gamma_2 nk_s k_c + 4c_2 mk_z$
k_f	$c_2\gamma_2 nk_s k_c(k_f - 1) + c_2 mk_z \log_2(k_f)$

Fig. 7. Complexity for the PMPDDP tree computation shown for the number of clusters computed. The additional expense of computing the estimated scatter is not considered

scatter, it is necessary to compute the principal direction of all the leaf clusters before we split them. We effectively incur the expense of computing an additional level in the PDDP tree, which doubles the number of splits computed. Therefore, the actual cost of computing a PDDP clustering of **CZ** when using the estimated scatter is

$$cost_{\texttt{clusterCZ}} = c_2\gamma_2 nk_s k_c(2k_f - 1) + c_2 mk_z \log_2(2k_f). \tag{24}$$

For clarity, we reproduce all the costs of computing a PMPDDP clustering in Table 3, and all the assumptions used to write the formulas in Fig. 8. Note that the costs are higher than computing a PDDP clustering. This is expected since we already incur more cost than a PDDP clustering just by obtaining the section representatives that comprise **C**, assuming $k_c > k_f$.

6.1 Complexity for One Varying Parameter

In this section we produce the PMPDDP complexity formulas for the case in which we vary the number of representatives k_z used to approximate each data item and the number of representatives k_c produced for each section of data, while leaving all other parameters fixed. We also produce formulas for the cost of PMPDDP with respect to the number of data items m and the number of attributes n.

Before we proceed further, we collect the results from the formulas in (13, 15, 24) and write the total cost of PMPDDP as:

$$m\bigg(c_1\gamma_1 n \log_2(k_c) + \gamma_2 nk_c + k_c k_z + \gamma_2 nk_z^2$$

$$+ \frac{1}{3}k_z^3 + c_2\gamma_2 \frac{n}{m}k_s k_c(2k_f - 1) + c_2 k_z \log_2(2k_f)\bigg). \tag{25}$$

We use this result when computing the formulas for each instance.

Table 3. Collected costs of PMPDDP, including the costs of obtaining the LMFR. See Table 2 for a definition of the parameters, and Fig. 8 for the assumptions made when writing the formulas

Operation	Amortized cost
Clustering sections to obtain **C**	$c_1\gamma_1 nm \log_2(k_c)$
Find distance from data points to centers	$\gamma_2 nmk_c$
Find k_z closest centers	$mk_c k_z$
Compute best least-squares approximation	$m(\gamma_2 nk_z^2 + \frac{1}{3}k_z^3)$
Cluster the representation **CZ** using PDDP	$c_2\gamma_2 nk_s k_c(2k_f - 1) + c_2 mk_z \log_2(2k_f)$
Compare cost of PDDP	$c_1\gamma_1 nm \log_2(k_f)$

> **Obtaining C**:
> 1. PDDP is used to obtain the section representatives
> 2. A perfectly balanced binary tree is created
> 3. each section has the same value of k_d and k_c
> 4. $k_d k_s = m$
> 5. k_c is an integer power of two
>
> **Obtaining Z**:
> 6. $k_z \ll k_c$, so direct search for k_z closest representatives in **C** is less expensive than sorting
> 7. same value for k_z is used for all sections
> 8. normal equations are used to obtain least-squares
> 9. additional $O(k_z^2)$ term from least squares is ignored
>
> **Clustering CZ**:
> 11. PDDP is used to obtain the clustering
> 12. A perfectly balanced binary tree is created
> 13. Scatter is estimated by pre-computing the splits for all leaves
> 14. k_f is an integer power of 2

Fig. 8. Summary of the assumptions used to obtain the formulas in Table 3

Varying k_z

In this section we show the cost of PMPDDP when all parameters except k_z are fixed. This will demonstrate the effect on the overall cost of PMPDDP when changing the number of representatives used to approximate each data item. Since k_z is independent from all other parameters, it is possible to fix the remaining parameters to constant values.

We start by examining (25) and extracting only those components that depend on k_z. Note that while the other components may contribute a very high cost, that cost will be fixed. The resulting formula is

$$m\left((k_c + c_2 \log_2 (2k_f)) k_z + \gamma_2 n k_z^2 + \frac{1}{3} k_z^3 \right). \tag{26}$$

This formula indicates that there may be a strong linear component in k_z when the quantity $\gamma_2 n k_z^2$ is relatively small, as would probably be the case for relatively low-dimension dense data sets. In the general case, since we expect k_z to be small, the square term will dominate the costs. With k_z sufficiently large, the cost will grow cubically.

Increasing k_z is expensive from a memory standpoint, since each column of **Z** has k_z nonzero entries. Keeping k_z small controls both the computation cost and the memory footprint of the LMFR.

Varying k_c

The other PMPDDP parameter we consider is the number of representatives per section k_c. We demonstrated in [317] that k_c is important to clustering accuracy, so it is useful to know the trade-off in cost. As before, if we only consider the elements of formula (25) that involve k_c, we have the result

$$m \left(c_1 \gamma_1 n \log_2 (k_c) + \gamma_2 n k_c + k_c k_z + c_2 \gamma_2 \frac{n}{m} k_s k_c (2k_f - 1) \right).$$

If we factor out the k_c term, we have the result

$$m \left(c_1 \gamma_1 n \log_2 (k_c) + k_c \left(\gamma_2 n + k_z + c_2 \gamma_2 \frac{n}{m} k_s (2k_f - 1) \right) \right). \qquad (27)$$

We expect that the cost of PMPDDP will increase slightly more than linearly with k_c due to the logarithmic term.

Varying n

We now consider the contribution to the cost from the number of attributes n. Taking all the terms from (25) that involve n, we have

$$m \left(c_1 \gamma_1 n \log_2 (k_c) + \gamma_2 n k_c + \gamma_2 n k_z^2 + c_2 \gamma_2 \frac{n}{m} k_s k_c (2k_f - 1) \right).$$

We can factor n from this formula with the resulting cost being

$$nm \left(c_1 \gamma_1 \log_2 (k_c) + \gamma_2 k_c + \gamma_2 k_z^2 + c_2 \gamma_2 \frac{1}{m} k_s k_c (2k_f - 1) \right). \qquad (28)$$

From this result, we expect the cost of PMPDDP to be linear in the number of attributes.

Varying m

The final result we consider is the cost in terms of the number of data items m. Note that in (25), all terms inside the outermost parenthesis are dependent on m except

$$c_2 \gamma_2 \frac{n}{m} k_s k_c (2k_f - 1),$$

since this term will be multiplied by m. With this consideration, and with all values fixed except m, and rather than rewriting the entire formula, we can recast (25) as

$$c_3 m + c_4, \qquad (29)$$

where c_3 and c_4 encompass the appropriate parameters in (25). From this result we can expect that PMPDDP is linear in the number of data items m.

7 Experimental Results

In this section we show some experimental results for the PMPDDP clustering method for both a dense and a sparse large data set. We subsampled each data set so that we could directly compare the results of a PDDP clustering with a PMPDDP clustering. We compare the quality of the two clusterings using scatter and entropy as measures.

Recall from Sect. 5.1 that PMPDDP uses an estimated scatter value to determine the cluster that is split next. To determine the effect on clustering quality of using the estimated scatter, we include results for clustering quality using the computed scatter. Computing the scatter required that the data be reconstructed. To minimize the amount of additional memory, we reconstructed 50 data points at a time. When we used the computed scatter to select the cluster to split next, we did not precompute the splits of the leaf clusters.

The algorithms were implemented in MATLAB and the experiments were performed on an AMD XP2200+ computer with 1 GB of memory and 1.5 GB of swap space.

7.1 Data Sets

We used two data sets to evaluate the method: one dense and one sparse. The dense data set was the KDD Cup 1998 data set [72], which consists of network connection data. Since the data set was designed to test classification algorithms, it was labeled. Connection types were either "normal" or some kind of attack. We combined both the training and the test data into one large data set. Categorical attributes were converted to binary attributes as needed. Each attribute was scaled to have a mean of 0 and a variance of 1. Post-processing, the entire data set occupied over 4 GB of memory.

The sparse data set was downloaded from the Web. Topics were selected so that a Google search on a given topic would return at least 200 hits. We assumed that the top 200 documents returned were relevant to the search topic. Each Web page was treated as a document. Any word that only appeared in one document was removed from the dictionary, as were all stop words. The words were also stemmed using Porter's stemming algorithm. Each document vector was scaled to unit length.

The data sets are summarized in Table 4. A more in-depth description of how the data were processed is available in [317].

7.2 Performance Measures

We used two different performance measures to evaluate the comparative quality of the clustering. Those measures were scatter and entropy.

The scatter s_C of a cluster $\mathbf{M}_C$ is defined as:

$$s_C \overset{\text{def}}{=} \sum_{j \in C} (\mathbf{x}_j - \mathbf{w}_C)^2 = \|\mathbf{M}_C - \mathbf{w}_C \mathbf{e}^T\|_F^2, \qquad (30)$$

Table 4. Summary of the datasets used in the experiments. The KDD data set is the KDD Cup 1998 data set from the UCI:KDD machine learning repository [72], and the Web data were produced at the University of Minnesota

Dataset	KDD	Web
Number of samples (m)	4,898,431	25,508
Number of attributes (n)	122	733,449
Number of categories	23	1,733

where $\mathbf{w}_C$ is the mean of the cluster, $\mathbf{e}$ is the m-dimensional vector $[1\ 1\ \cdots\ 1]^T$, and $\|\ \|_F$ is the Frobenius norm. For some $n \times m$ matrix $\mathbf{A}$, the Frobenius norm of $\mathbf{A}$ is

$$\|A\|_F = \sqrt{\sum_{\substack{1 \le i \le n \\ 1 \le j \le m}} a_{i,j}^2}, \tag{31}$$

where $a_{i,j}$ is the entry in the ith row and jth column of $\mathbf{A}$. A low scatter value indicates good cluster quality. Since scatter is a relative performance measure, it only makes sense to use the scatter to compare clusterings having the same cardinality.

The entropy e_j of cluster j is defined by:

$$e_j \stackrel{\text{def}}{=} -\sum_i \left(\frac{c(i,j)}{\sum_i c(i,j)} \right) \cdot \log \left(\frac{c(i,j)}{\sum_i c(i,j)} \right), \tag{32}$$

where $c(i,j)$ is the number of times label i occurs in cluster j. If all the labels of the items in a given cluster are the same, then the entropy of that cluster is 0. Otherwise, the entropy is positive. The total entropy for a given clustering is the weighted average of the cluster entropies:

$$e_{\text{total}} \stackrel{\text{def}}{=} \frac{1}{m} \sum_i e_i \cdot k_i. \tag{33}$$

The lower the entropy, the better the quality. As with the scatter, entropy is a relative performance measure, so the same caveats apply.

7.3 KDD Results

The KDD intrusion detection data were divided into 25 random samples. It was only possible to compute a PDDP clustering of the data up to a combination of the first 5 samples of data. After that, the amount of memory required for data and overhead exceeded the capacity of the computer.

The parameters and results for the KDD intrusion detection data set are summarized in Table 5. We show the results for the combinations through the first five pieces of data and the results for the entire data set. The clustering

Table 5. Comparison of a PDDP clustering with a PMPDDP clustering of the KDD data for various sample sizes. It was not possible to compute a PDDP clustering past the largest sample size shown since the data would not fit into memory. It was not possible to compute the scatter for PMPDDP for the entire data set, since it would not fit into memory. Also included are results for a modified PMPDDP clustering method, which uses the computed scatter (c.s.) rather than the estimated scatter. See Table 2 for a definition of the parameters, and Sect. 7.1 for a description of the data

Dataset	KDD					
m	195,937	391,874	587,811	783,748	979,685	4,898,431
k_s	5	10	15	20	25	125
k_c	392	392	392	392	392	392
k_z	3	3	3	3	3	3
k_f	36	36	36	36	36	36

Normalized scatter values, lower is better						
PDDP	$3.179e-04$	$3.279e-04$	$3.276e-04$	$3.290e-04$	$3.288e-04$	Will not fit
PMPDDP	$3.257e-04$	$3.236e-04$	$3.271e-04$	$3.250e-04$	$3.275e-04$	N.A.
PMPDDP c.s.	$3.271e-04$	$3.258e-04$	$3.250e-04$	$3.245e-04$	$3.255e-04$	N.A.

Entropy values, lower is better						
PDDP	0.127	0.130	0.129	0.124	0.120	Will not fit

PMPDDP	0.0590	0.0585	0.0546	0.120	0.114	0.113
PMPDDP c.s.	0.125	0.127	0.126	0.112	0.120	0.113

Time taken by experiments, in seconds, on XP 2200+						
PDDP	39.72	89.68	140.45	204.87	282.44	Will not fit
Compute **CZ**	216.66	450.41	674.49	872.15	1,108.56	5,652.59
Cluster **CZ**	15.63	32.71	52.67	69.07	89.62	492.81
Cluster **CZ** c.s.	20.20	45.18	73.59	96.42	120.69	447.36
PMPDDP totals	232.29	483.12	727.16	941.22	1,198.18	6,145.40

Memory occupied by representation, in MB						
M	191.2	382.4	573.6	764.9	956.2	**4,780**
CZ	8.24	16.48	24.72	39.00	48.75	**206**

quality from PMPDDP is comparable to PDDP in both the scatter and the entropy performance measures. The memory savings are significant. The costs are higher for PMPDDP, but the majority of the time is spent computing the factored representation of the data. Once the factored representation is available, clusterings of different sizes can be computed relatively inexpensively.

Examining the results for the PMPDDP clustering using the computed scatter as compared to PMPDDP using the estimated scatter, we can see that the scatter values are similar for the two approaches. The entropy is better for the smaller sample sizes when using the estimated scatter, but this could be a result of some bias in the data set. A few labels are assigned to most of the data points, while the remaining labels are assigned to relatively few data points. A small number of data points can move and change the entropy results significantly. In any case, once more data are present, the advantage in entropy values for the estimated scatter is no longer present. As such, this anomaly probably does not represent any significant advantage in general to using the estimated scatter. Also, we can see that the clustering times are close enough so that there is no advantage in speed when using the estimated scatter for this dense data set.

The entire KDD intrusion detection data set in its original representation would occupy 4.78 GB of memory, beyond the limits of most desktop workstations. The factored representation of the data requires only about 206 MB of memory for the PMPDDP parameters selected, leaving plenty of memory space for clustering computation overhead on even a 512 MB workstation.

The time taken as the number of data items increased is shown in Fig. 9. For this data set, PMPDDP costs scale linearly with the number of data items. This agrees with the complexity analysis in Sect. 6.1.

7.4 Web

The Web data were divided into eight random samples. It was not possible to compute a PDDP clustering for more than a combination of six samples of the data, since after that point the program terminated abnormally due to lack of swap space.

The parameters and results are shown in Table 6, including the results for the entire data set. The PMPDDP and PDDP clusterings again have similar quality with respect to the scatter and entropy. Note that the memory savings

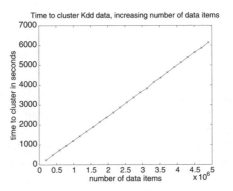

Fig. 9. Time taken for a PMPDDP clustering of the KDD data set with an increasing number of data items. The parameters used in the experiments are shown in Table 5

Table 6. Comparison of a PDDP clustering and a PMPDDP clustering of the web data set for various sample sizes. The entire data set would not fit into memory, so it was not possible to perform a PDDP clustering for the entire data set. Since the data would not fit, it was not possible to compute the scatter of PMPDDP for the entire data set. Also included are results for a modified PMPDDP clustering method, which uses the computer scatter (c.s.) rather than the estimated scatter. See Table 2 for the parameter definitions and Sect. 7.1 for a description of the data

Dataset	Web						
m	40,688	81,376	122,064	162,752	203,440	244,128	325,508
k_s	5	10	15	20	25	30	40
k_c	81	81	81	81	81	81	81
k_z	3	3	3	3	3	3	3
k_f	200	200	200	200	200	200	200

Normalized scatter values, lower is better							
PDDP	0.7738	0.7760	0.7767	0.7778	0.7789	0.7787	Will not fit
PMPDDP	0.7762	0.7792	0.7802	0.7809	0.7819	0.7829	N.A.
PMPDDP c.s.	0.7758	0.7788	0.7796	0.7800	0.7812	0.7817	0.7820

Entropy values, lower is better							
PDDP	5.043	5.483	5.692	5.836	5.922	5.990	Will not fit
PMPDDP	5.168	5.624	5.843	6.004	6.094	6.175	6.869
PMPDDP c.s.	5.127	5.598	5.842	5.987	6.088	6.161	6.253

Time taken by experiments, in seconds, on an XP 2200+							
PDDP	1461	2527	3421	4359	5277	6286	Will not fit
Compute **CZ**	5909	11,783	17,648	23,508	29,414	35,288	47,058
Cluster **CZ**	9174	17,654	26,278	34,565	43,591	51,992	68,416
Cluster **CZ** c.s.	15,192	30,208	44,908	59,634	74,542	89,614	11,9762
PMPDDP total	15,083	29,437	43,926	58,073	73,005	87,279	115,475

Memory occupied by representation, in MB							
M	115.3	227.3	339.9	451.4	563.6	674.6	**897.3**
CZ	43.54	84.72	126.0	167.0	208.5	248.9	**330.3**

are not as significant as those for the dense data set. The sparse data already use a representation which saves a considerable amount of memory. Plus, the **C** matrix is considerably more dense than the original representation of the data. This is due to the fact that the **C** matrix comprises cluster centroids. A centroid of any given cluster must contain a word entry for any data point in the cluster, which has that word as an element. Therefore, the centroid of a cluster of sparse data is usually denser than any given item in the cluster. The higher density also accounts for some of the additional expense incurred during clustering, since the greater density is associated with an increase in the number of multiplications required to obtain the principal direction of the cluster.

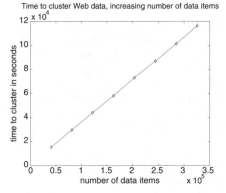

Fig. 10. Time taken for a PMPDDP clustering of the web data with an increasing number of data items. The results are a graphical representation of the times from Table 6

It was not possible to cluster the entire data set using the original representation of the data, which occupied 897 MB of memory, while the LMFR at 330 MB left sufficient memory space for clustering overhead.

The comparison between clustering quality and cost between standard PMPDDP and PMPDDP using the computer scatter is much more pronounced than for the dense data. Using the estimated scatter saves a significant amount of time during the clustering process, even though it requires computing twice the number of splits. The scatter values when clustering using the computed scatter are slightly better than those for the estimated scatter. The entropy values are better as well. However, the amount of time saved when using the estimated scatter is enough so that we would still recommend using it over the computed scatter.

The time taken as the number of data items increased is shown in Fig. 10. As with the KDD data set, the time taken to compute a complete PMPDDP clustering of the web data is linear in the number of data items. This agrees with the complexity analysis in Sect. 6.1.

8 How to Apply PMPDDP in Practice

We realize that while the experiments demonstrate that PMPDDP is a scalable extension of PDDP, they do not give much intuition on how to select the parameters. There are much more extensive experiments in [317], and using those results, we can provide some guidelines on how to apply PMPDDP in practice.

The LMFR construction is the point where the parameter choices are most relevant. Results indicate that even for an LMFR with the exact same memory footprint, minimizing the number of sections is beneficial. In other words, it is best to process the original data in pieces which are as large as can be fit into memory while still allowing room for overhead and room for the new matrices being created. Realize that it is not necessary for the sections to have the same size, so if there one some remainder data, they can be used to construct a smaller section, with the remaining parameters adjusted accordingly.

There is a trade-off when selecting the values of k_c and k_z, since those are the two user-selectable parameters that control the size of the LMFR. The experimental results indicated that increasing k_z to numbers above 7 or 8 does not increase clustering quality, and values of 3–5 provide good results in most cases.

The single most important parameter is k_c, the number of representatives produced per section. Both the accuracy of the clustering and the approximation accuracy of the LMFR are strongly influenced by the value of k_c. Therefore, k_c should be chosen such that it will maximize the memory footprint of the global LMFR.

Using the above information, we recommend starting with a small value for k_z and then selecting k_c to maximize memory use. If the data set in question is sparse, it would be advisable to test the parameter choice on a small piece of data to determine the increase in density, so that some estimate of the final memory footprint of the LMFR could be obtained.

Note that applying the above technique may produce an LMFR, which is much larger than the size necessary to obtain good clustering quality. For instance, the experimental results in Sect. 7.3 were good with an LMFR that did not occupy as much memory as possible. However, it is difficult to know beforehand how much reduction a given data set can tolerate. The optimum memory reduction, if there is such a result, would be strongly data dependent. Therefore, it is difficult to do other than suggest making the LMFR as large and, correspondingly, as accurate a representation of the original data as available memory allows.

9 Conclusions

In this chapter we presented a method to extend the PDDP clustering method to data sets that cannot fit into memory at once. To accomplish this, we construct an LMFR of the data. The LMFR transparently replaces the original representation of the data in the PDDP method. We call the combination of constructing an LMFR and clustering it using PDDP PMPDDP.

The LMFR is comprised of two matrices. The LMFR is computed incrementally using relatively small pieces of the original data called sections. Each section of data is clustered, and the centroids of the clusters form the first matrix $\mathbf{C}_j$. The centroids are then used to construct a least-squares approximation to each data point. The data loadings from the least-squares approximation are used to construct the second matrix $\mathbf{Z}_j$. Memory is saved since only a small number of centroids are used to approximate each data item, making $\mathbf{Z}$ very sparse. $\mathbf{Z}$ must be represented in a sparse matrix format in order to realize the memory savings. Once a $\mathbf{C}_j$ and a $\mathbf{Z}_j$ are available for each section, they are assembled into a global representation of all the data, $\mathbf{CZ}$. The matrices $\mathbf{CZ}$ can then be used in place of the original representation of the data. The product $\mathbf{CZ}$ is never computed explicitly, since the product would take up at least as much space as the original data. Unlike many other approximating techniques, the LMFR provides a unique representation of each data item.

We provided a complexity analysis of the cost of constructing the LMFR. This provides a useful guide when determining how to choose the parameters if the time of computation is critical. In the process, we showed that PDDP is theoretically linear in the number of data items and the number of attributes, which has been shown to be the case experimentally.

We then described the PMPDDP clustering algorithm. PMPDDP uses the LMFR in place of the original data to obtain a clustering of the data. Since each original data item has a corresponding column in the LMFR, mapping the clustering of the LMFR to the original data is trivial. Therefore, a clustering of the LMFR is a clustering of the original data.

PDDP is uniquely suited to clustering the LMFR since PDDP does not require similarity measures to determine the clustering. Therefore, no data need to be reconstructed and no full or even partial products of $\mathbf{CZ}$ are computed. To avoid reconstructing the data, an estimate of the scatter is used in place of the computed scatter when determining the cluster in the PDDP tree that is split next.

With the complexity analysis, we showed PMPDDP is linear in the number of data items and the number of attributes. Thus, PMPDDP extends PDDP to large data sets while remaining scalable.

Next, we provided some experimental results. The experiments demonstrated that it is possible to produce a PMPDDP clustering which has quality comparable to a PDDP clustering while saving a significant amount of memory. Therefore, it is possible to cluster much larger data sets than would otherwise be possible using the standard PDDP method. Additional data sets were shown to be clustered successfully using PMPDDP in [317].

We also showed the effect of replacing the estimated scatter, as used in PMPDDP, with the computed scatter, when determining the cluster to be split next. For the dense data set, the difference was neutral with respect to clustering quality and clustering time. However, for the sparse data set, a significant amount of time can be saved during clustering by using the estimated scatter. Clustering quality was slightly inferior, but the reductions in clustering times more than made up for the differences in clustering quality.

Finally, we described how we would expect PMPDDP would be applied by a casual user. While we cannot guarantee that our suggestions provide an optimal balance between memory used and clustering quality, we believe that using our suggestions would provide the best clustering quality obtainable considering the amount of memory available on the workstation being used.

Acknowledgment

This work was partially supported by NSF grant 0208621.

Clustering with Entropy-Like k-Means Algorithms

M. Teboulle, P. Berkhin, I. Dhillon, Y. Guan, and J. Kogan

Summary. The aim of this chapter is to demonstrate that many results attributed to the classical k-means clustering algorithm with the squared Euclidean distance can be extended to many other distance-like functions. We focus on entropy-like distances based on Bregman [88] and Csiszar [119] divergences, which have previously been shown to be useful in various optimization and clustering contexts. Further, the chapter reviews various versions of the classical k-means and BIRCH clustering algorithms with squared Euclidean distance and considers modifications of these algorithms with the proposed families of distance-like functions. Numerical experiments with some of these modifications are reported.

1 Introduction and Motivation

The problem of clustering can be briefly described as follows: partition a given collection of objects into disjoint subcollections (called clusters) so that the objects in a cluster are more similar to each other than to those in other clusters. The number of clusters, often denoted by k, also has to be determined. Clustering techniques are used to discover natural groups in data sets, and to identify abstract structures that might reside there, without having any background knowledge of the characteristics of the data. They have been used in a variety of areas including: computer vision, VLSI design, data mining, text mining, bioinformatics, Web mining, and gene expression analysis. The importance and interdisciplinary nature of clustering is evident through its vast literature. For a recent state-of-the-art review on clustering, see, e.g., Jain et al. [245], Mishra and Motwani [339], and Berkhin in Chap. 2 in this book and references therein.

A large class of clustering techniques attempt to partition the data while optimizing an associated objective function. A common feature of the most basic optimization formulation of clustering is that the problem is nonconvex and often nonsmooth, thus falling into one of the most difficult areas of the optimization field. A popular clustering technique is the k-means algorithm and many of its modifications and refinements (see, for example, [131, 136, 144, 166, 277, 323]). In fact, this algorithm is a local search optimization-type algorithm, or, more precisely, a gradient-type method (see, e.g., [67]) when appropriately formulated.

The aim of this chapter is to demonstrate that many results attributed to the classical k-means clustering algorithm with the squared Euclidean distance can be extended with many other distance-like functions, see Sect. 2 and Sect. 3. A step in this direction has already been made by, for example, Tishby et al. [421], Dhillon et al. [134], Kogan et al. [279], Kogan et al. [280, 282], and Banerjee et al. [46, 47].

We focus on entropy-like distances based on Bregman [88] and Csiszar [119] divergences, which have already been shown to be quite useful in various optimization contexts (see, for instance, Bregman [88], Censor and Lent [101], Censor and Zenios [102], and references therein for the former, and Teboulle [415–417] and references therein for the latter). The lack of symmetry with respect to the vector arguments of these distance-like functions (as opposed to the classical squared Euclidean distance) allows for more flexibility in the formulation of clustering problems. A recent interesting study in that direction is the work of Banerjee et al. [46, 47] where the authors have derived the cluster "centroids" by minimizing a sum of distances based on Bregman divergences, but with respect to the second argument of the Bregman distance function. This allows the authors to produce a remarkable result despite the nonconvexity of the Bregman distance with respect to its second argument, see Sect. 4 for details. In this chapter we treat the analysis of the minimization problem with respect to either one of the vector arguments, pointing out the differences/similarities in the resulting derived centroids. This allows one to consider k-means clustering algorithms within a unified framework and paves the way for the development of new clustering algorithms. In particular, these algorithms are capable of handling non-negative data with a wide range of choices for the distance measure to accommodate a variety of specific problems. This leads to several new types of cluster centers (or centroids), see Sect. 4.

In Sect. 5 we concentrate on the batch version of the algorithm [166], the incremental version [144], and the combination of these two algorithms [219, 277, 458]. The section also discusses a strategy to handle large datasets by cutting an original dataset into small dense clusters and handling each small cluster as a single "cluster-point." This approach leads to a significant reduction of the dataset size. The idea is borrowed from [459] (which uses the squared Euclidean distance), and we call the resulting algorithm the "BIRCH type" algorithm. Experimental results demonstrating the usefulness of our approach for various datasets and choices of entropy-like distances are given in Sect. 6.

2 An Optimization Formulation of Clustering Problems

There are several ways to formulate a clustering problem. We start with one of the most basic formulation, which allows us to set up notations and terminology.

Let $\mathcal{A} = \{a^1, \ldots, a^m\}$ be a set of points in a subset S of an n-dimensional Euclidean space $\mathbb{R}^n$ (i.e., $a^i = (a^i_1, \ldots, a^i_n)^T$, $i = 1, \ldots, m$). Consider a partition $\Pi = \{\pi_1, \ldots, \pi_k\}$ of the set, i.e.,

$$\pi_1 \cup \cdots \cup \pi_k = \{a^1, \ldots, a^m\}, \ \pi_i \subseteq \mathcal{A}, \text{ and } \pi_i \cap \pi_j = \emptyset \text{ if } i \neq j.$$

Given a real valued function q whose domain is a subset of $\{a^1, \ldots, a^m\}$, the quality of the partition Π is given by $Q(\Pi) = q(\pi_1) + \cdots + q(\pi_k)$. The problem is to identify an optimal partition $\Pi^\circ = \pi_1^\circ \cup \cdots \cup \pi_k^\circ$, i.e., the one that minimizes $Q(\Pi) = q(\pi_1) + \cdots + q(\pi_k)$. Often the function q is associated with a "dissimilarity measure," or a *distance-like* function $d(u, v)$ that satisfies the following basic properties:

$$d(u, v) \geq 0 \ \forall (u, v) \in S \text{ and } d(u, v) = 0 \iff u = v.$$

We will call $d(\cdot, \cdot)$ a *distance-like* function, since we are not necessarily asking for d to be either symmetric or to satisfy the triangle inequality.

To describe the relation between q and d we define a centroid x of a cluster π by

$$x = x(\pi) = \arg \min_{y \in S \subset \mathbb{R}^n} \sum_{a \in \pi} d(y, a). \tag{1}$$

Note that since $d(\cdot, \cdot)$ is not necessarily symmetric, by reversing the order of the variables we could also define a centroid by minimizing with respect to the second argument; this i.e., further discussed in Sect. 4.

If $q(\pi)$ is defined as $\sum_{a \in \pi} d(x(\pi), a)$, then centroids and partitions are associated as follows:

1. For each set of k centroids $\{x^1, \ldots, x^k\}$ one can define a partition $\{\pi_1, \ldots, \pi_k\}$ of the set $\{a^1, \ldots, a^m\}$ by:

$$\pi_i = \{a^j \ : \ d(x^i, a^j) \leq d(x^l, a^j) \text{ for each } l \neq j\}$$

   (break ties arbitrary).
2. Given a partition $\{\pi_1, \ldots, \pi_k\}$ of the set $\{a^1, \ldots, a^m\}$, one can define the corresponding centroids $\{x^1, \ldots, x^k\}$ by:

$$x^i = x(\pi_i) = \arg \min_{x \in S} \sum_{a \in \pi_i} d(x, a). \tag{2}$$

Note that the spherical k-means clustering algorithm introduced by Dhillon and Modha [136] is covered by this model. When the minimization problem in (2) is convex, the above algorithm has been coined as the *convex k-means* algorithm in [341].

While the discrete clustering minimization problem

$$\min Q(\Pi) \text{ where } Q(\Pi) = \sum_{i=1}^{k} q(\pi_i) = \sum_{i=1}^{k} \sum_{a \in \pi_i} d(x^i, a)$$

is stated here in terms of partitions, one should keep in mind that a partition is always associated with a set of centroids. In the beginning of the chapter we concentrate on centroids (see Sect. 5 where we focus on partitions) and state the continuous clustering minimization problem as follows:

$$(C) \qquad \min_{x^1, \ldots, x^k \in S} f(x^1, \ldots, x^k) \text{ where } f(x^1, \ldots, x^k) = \sum_{i=1}^{m} \min_{1 \le l \le k} d(x^l, a^i).$$

The nonsmooth optimization formulation (C) of the k-means clustering problem leads to an algorithm [418] that recovers the deterministic annealing procedure introduced by Ross et al. [373].

We now indicate two of the most common and popular choices of d. The first choice is with $S = \mathbb{R}^n$ and d being an arbitrary norm $\| \cdot \|$ in $\mathbb{R}^n$, in which case the objective function of problem (C), which for convenience will be called (C_1), becomes

$$(C_1) \qquad \min_{x^1, \ldots, x^k \in \mathbb{R}^n} f(x^1, \ldots, x^k) \text{ where } f(x^1, \ldots, x^k) = \sum_{i=1}^{m} \min_{1 \le l \le k} \|x^l - a^i\|.$$

The second popular choice of d is the squared Euclidean norm, $S = \mathbb{R}^n$, that leads to problem (C_2):

$$(C_2) \qquad \min_{x^1, \ldots, x^k \in \mathbb{R}^n} f(x^1, \ldots, x^k) \text{ where } f(x^1, \ldots, x^k) = \sum_{i=1}^{m} \min_{1 \le l \le k} \|x^l - a^i\|^2.$$

The problem (C_1) with an L_1 norm and the problem (C_2) with a squared L_2 norm lead to two of the most popular clustering algorithms called the k-median algorithm and the k-means algorithm, respectively (see, e.g., [87]).

In Sect. 3 we recall some basic results and definitions of distance-like functions.

3 Distance-Like Functions

A specific choice of a distance-like function is dictated by an application at hand. Often the data set resides in a prescribed closed convex subset S of

$\mathbb{R}^n$, most often S is the non-negative octant, or the product of real intervals (see, e.g., [60, 134, 136, 393]). To handle such situations, we focus on two well-established distance-like functions based on φ-divergences and Bregman distances. These are generated by the choice of an appropriate convex kernel function. For a particular choice of kernels, both measures coincide with the so-called Kullback–Leibler (KL) divergence or relative entropy. Accordingly, we call these *Entropy-like distances*. The purpose of this section is to provide a brief review on these objects and recall some of their basic and useful properties. Our terminology uses standard notions of convex analysis that can be found in the classical book of Rockafellar [372] (for notations/results not explicitly given here, the reader is referred to [372]).

3.1 φ-Divergences

Consider a collection of vectors $\{a^1, \ldots, a^m\}$ in the n-dimensional non-negative octant $S = \mathbb{R}_+^n$. A natural "distance" between vectors is provided by the so-called class of φ-divergence measures, which was introduced by Csiszar [119] as a generalized measure of information on the set of probability distributions to measure the closeness between two probability measures. That is, Csiszar divergences generalize the concept of relative entropy. For more results on convex statistical distances see, for instance, the monograph [315] and references therein. Here we need not restrict ourselves to the probability setting (as is often the case in Information Retrieval), and we consider the more general case of arbitrary vectors in the non-negative octant (our presentation follows Teboulle [415–417], where more results and details on using φ-divergences in various contexts can be found).

First we define the class of functions from which the distance-like functions will be derived. Let $\varphi : \mathbb{R} \to (-\infty, +\infty]$ be a proper closed (lower semicontinuous, for short l.s.c.) function.[1] We denote its effective domain by

$$\text{dom } \varphi := \{t : \varphi(t) < +\infty\}.$$

We further assume that dom $\varphi \subseteq [0, +\infty)$, $\varphi(t) = +\infty$ when $t < 0$ and φ satisfies the following:
(i) φ is twice continuously differentiable on int(dom φ) $= (0, +\infty)$,
(ii) φ is strictly convex on its domain,
(iii) $\lim_{t \to 0^+} \varphi'(t) = -\infty$,
(iv) $\varphi(1) = \varphi'(1) = 0$ and $\varphi''(1) > 0$.

---

[1] Recall that φ is called proper if $\varphi(t) > -\infty$ for all t and dom $\varphi \neq \emptyset$, and l.s.c. (closed) if $\liminf_{s \to t} \varphi(s) = \varphi(t)$ (the epigraph of φ is a closed set).

We denote by Φ the class of functions satisfying (i)–(iv). Given $\varphi \in \Phi$, for $x, y \in \mathbb{R}^n$ we define $d_\varphi(x, y)$ by

$$d_\varphi(x, y) = \sum_{j=1}^{n} y_j \varphi(x_j / y_j). \tag{3}$$

A few words about these specific assumptions are in order. Since we will be interested in minimization problems involving d_φ, assumptions (i)–(ii) ensure the existence of global minimizers, assumption (iii) enforces the minimizer to stay in the positive octant, and finally (iv) is just a normalization, allowing us to handle vectors in $\mathbb{R}^n_+$ (rather than probabilities).

The functional d_φ enjoys the required basic property of a distance-like function, namely one has:

$$\forall (x, y) \in \mathbb{R}^n \times \mathbb{R}^n \quad d_\varphi(x, y) \geq 0 \text{ and } d_\varphi(x, y) = 0 \text{ iff } x = y.$$

Indeed, the strict convexity of φ and (iv) imply:

$$\varphi(t) \geq 0, \text{ and } \varphi(t) = 0 \text{ iff } t = 1.$$

Moreover, it can be easily verified that d_φ is *jointly convex* in (x, y).

Example 1. Typical interesting examples of φ include:

$$\varphi_1(t) = t \log t - t + 1, \text{ dom } \varphi = [0, +\infty).$$

$$\varphi_2(t) = -\log t + t - 1, \text{ dom } \varphi = (0, +\infty).$$

$$\varphi_3(t) = (\sqrt{t} - 1)^2, \text{ dom } \varphi = [0, +\infty).$$

The first example leads to the so-called relative entropy type distance of KL defined on $\mathbb{R}^n_+ \times \mathbb{R}^n_{++}$ by

$$d_{\varphi_1}(x, y) \equiv KL(x, y) := \sum_{j=1}^{n} x_j \log \frac{x_j}{y_j} + y_j - x_j. \tag{4}$$

Note that for unit L_1 norm vectors x and y with non-negative coordinates $\sum_{j=1}^{n} x_j = \sum_{j=1}^{n} y_j = 1$, and the function $d_{\varphi_1}(x, y)$ becomes the standard KL divergence. Clustering results with the standard KL divergence have been recently reported by Berkhin and Becher [58], and Dhillon et al. [134].

By adopting the convention $0 \log 0 \equiv 0$, the KL functional can be continuously extended on $\mathbb{R}^n_+ \times \mathbb{R}^n_{++}$, i.e., it admits vectors with zero entries in its first argument (in text mining applications $x_j = y_j = 0$ corresponds to the case when a word/term does not occur in two documents and motivates the convention $0 \log (0/0) \equiv 0$). It turns out that KL remains one of the most fundamental distances, even when working and analyzing problems involving φ-divergences (see [417] for more details and results, and also Sect. 3.2).

The second example φ_2 also yields the KL distance but with reversed order of variables, i.e.,

$$d_{\varphi_2}(x, y) \equiv KL(y, x) = \sum_{j=1}^{n} y_j \log \frac{y_j}{x_j} + x_j - y_j, \qquad (5)$$

and the third choice gives the so-called Hellinger distance:

$$d_{\varphi_3}(x, y) = \sum_{j=1}^{n} (\sqrt{x_j} - \sqrt{y_j})^2$$

(additional examples can be found in [416]).

Note that the squared Euclidean distance cannot be recovered through the φ-divergence. Motivated by the analysis of a class of optimization algorithms in a recent work, Auslender et al. [35] have suggested a homogeneous functional $\hat{d}_{\varphi}$ of order 2 defined on $\mathbb{R}_{++}^n \times \mathbb{R}_{++}^n$ by:

$$\hat{d}_{\varphi}(x, y) = \sum_{j=1}^{n} y_j^2 \varphi(x_j / y_j). \qquad (6)$$

A particularly useful example of $\hat{d}_{\varphi}$ is provided with the choice

$$\varphi_4(t) = \frac{\nu}{2}(t - 1)^2 + \mu(t - \log t - 1), \ \text{dom} \ \varphi = (0, +\infty), \qquad (7)$$

where ν, μ are some positive parameters (see Auslender and Teboulle [34] for properties and applications). It leads to an interesting combination of the classical squared Euclidean distance with an additional new term, which can be interpreted as a "penalization" for positive data, through the log term. When $\mu \to 0$ the squared Euclidean distance is recovered.

3.2 Bregman Distance

Let $\psi : \mathbb{R}^n \to (-\infty, +\infty]$ be a closed proper convex function. Suppose that ψ is continuously differentiable on int(dom ψ) $\neq \emptyset$. The Bregman distance (also called "Bregman divergence") $D_\psi :$ dom $\psi \times$ int(dom $\psi) \to \mathbb{R}_+$ is defined by

$$D_\psi(x, y) = \psi(x) - \psi(y) - \langle x - y, \nabla\psi(y) \rangle, \qquad (8)$$

where $\langle \cdot, \cdot \rangle$ denotes the inner product in $\mathbb{R}^n$ and $\nabla\psi$ is the gradient of ψ.

This distance-like measure was introduced by Bregman [88]. It essentially measures the convexity of ψ, since $D_\psi(x, y) \geq 0$ if and only if the gradient inequality for ψ holds, i.e., if and only if ψ is convex. Thus, D_ψ provides a natural distance-like measure, and with ψ strictly convex one has $D_\psi(x, y) \geq 0$ and $D_\psi(x, y) = 0$ iff $x = y$.

Note that $D_\psi(x, y)$ is not a distance (it is not symmetric and it does not satisfy the triangle inequality). With the special choice dom $\psi = \mathbb{R}^n$ and $\psi(x) = 1/2\|x\|^2$, one obtains $D_\psi(x, y) = 1/2\|x - y\|^2$. More examples are given below.

In most applications, and in particular in optimization contexts, as has been already observed by Teboulle [415], the useful setting for Bregman distance is to consider convex functions of Legendre type, a concept introduced by Rockafellar [372]. Thus, we assume that the kernel function ψ is a closed proper strictly convex function, which is essentially smooth [372, p. 251]:

Definition 1. *A closed proper convex function* $\psi : \mathbb{R}^n \to (-\infty, +\infty]$ *is essentially smooth if it satisfies the following three conditions:*

(a) $C := int(dom\ \psi) \neq \emptyset$;
(b) ψ *is differentiable on* C;
(c) $\lim_{i \to \infty} \|\nabla\psi(x^i)\| = +\infty$, *for every sequence* $\{x^i\} \in C$ *converging to a boundary point* x *of* C.

Note that as for φ-divergences, the property (c) will enforce a minimizer of (1) with D_ψ as a distance-like function to stay in the interior of dom ψ. The class of functions such that C is an open convex set and ψ is a strictly convex function satisfying (a), (b), and (c) are called convex function of Legendre type [372, p. 258]. This class of functions is denoted by $\mathcal{L}(C)$.

We recall that the convex conjugate of $\psi : \mathbb{R}^n \to (-\infty + \infty]$ is defined by

$$\psi^*(y) = \sup_{x \in \mathbb{R}^n} \{\langle x, y \rangle - \psi(x)\}$$

and is a closed proper convex function.

A smooth convex function on $\mathbb{R}^n$ (i.e., finite and differentiable everywhere on $\mathbb{R}^n$) is in particular essentially smooth. From the results of [372, Sect. 26], it follows that for $\psi \in \mathcal{L}(C)$, the conjugate ψ^* of ψ is also in $\mathcal{L}(C^*)$, where $C^* = int(dom\ \psi^*)$ and the following useful relationships hold:

$$\nabla\psi^* = (\nabla\psi)^{-1}, \tag{9}$$

$$\psi^*(\nabla\psi(z)) = \langle z, \nabla\psi(z) \rangle - \psi(z). \tag{10}$$

These results help to compute centroids when distance functions are generated by Bregman divergences (see Sect. 4.3).

Example 2. With $C = \mathbb{R}^n_{++}$ and $\psi(x) := \sum_{i=1}^n x_i \log x_i - x_i$ (with the convention $0 \log 0 = 0$), we obtain the KL relative entropy distance

$$D_\psi(x, y) = \sum_{i=1}^{n} x_i \log \frac{x_i}{y_i} + y_i - x_i \ \forall\, (x, y) \in \mathbb{R}_+^n \times \mathbb{R}_{++}^n. \tag{11}$$

By reversing the order of variables in D_ψ, i.e.,

$$D_\psi(y, x) = \psi(y) - \psi(x) - \langle y - x, \nabla\psi(x) \rangle \tag{12}$$

(compare with (8)) using the kernel

$$\psi(x) = \frac{\nu}{2}\|x\|^2 + \mu \left[\sum_{i=1}^{n} x_i \log x_i - x_i \right], \tag{13}$$

we obtain

$$D_\psi(y, x) = \frac{\nu}{2}\|y - x\|^2 + \mu \sum_{i=1}^{n} \left[y_i \log \frac{y_i}{x_i} + x_i - y_i \right]. \tag{14}$$

More discussion on this example is presented in Sect. 4.

Example 3. Note that the choice $\psi(x) = -\sum_{i=1}^{n} \log x_i$ (known as the Burg entropy) leads to the distance

$$D_\psi(x, y) = \sum_{i=1}^{n} -\log \frac{x_i}{y_i} + \frac{x_i}{y_i} - 1.$$

We give now some other examples, taken from [415, p. 678].

Example 4. Most of the useful Bregman distances are *separable* (except for the trivial extension of the squared Euclidean distance on $\mathbb{R}^n$, which can be derived with the weighted norm $\psi(x) = x^T W x$, with W positive definite). Thus we let $K : \mathbb{R} \to (-\infty, +\infty]$ and

$$\psi(x) = \sum_{i=1}^{n} K(x_i). \tag{15}$$

In each of the examples below ψ is a closed proper strictly convex and essentially smooth function. In the first two examples ψ is cofinite, i.e., dom $\psi^* = \mathbb{R}^n$. In the last two examples one has $0 \in \mathrm{int}(\mathrm{dom}\ \psi^*)$. In each case, we give the conjugate function ψ^* and the corresponding induced Bregman distance. For the Cartesian product of n intervals (a, b) we use the notation $(a, b)^n$.

1. $K(t) = t \log t - t$ if $t \geq 0$, $K(t) = \infty$ if $t < 0$. Then [372, p. 105], $\psi^*(y) = \sum_{i=1}^{n} e^{y_i}$ and as already mentioned D_ψ recovers the KL relative entropy distance

$$D_\psi(x,y) = \sum_{i=1}^{n} x_i \log \frac{x_i}{y_i} + y_i - x_i \ \forall \ (x,y) \in \mathbb{R}_+^n \times \mathbb{R}_{++}^n. \qquad (16)$$

2. $K(t) = -\sqrt{1-t^2}$ if $|t| \leq 1$, $K(t) = \infty$ if $|t| > 1$. Then [372, p. 105], $\psi^*(y) = \sum_{i=1}^{n} \sqrt{1+y_i^2}$ and

$$D_\psi(x,y) = \sum_{i=1}^{n} \frac{1-x_i y_i}{\sqrt{1-y_i^2}} - \sqrt{1-x_i^2} \ \text{on} \ [-1,1]^n \times (-1,1)^n.$$

3. Burg entropy, $K(t) = -\log t$ if $t > 0$, $K(t) = \infty$ if $t \leq 0$. The conjugate function is given by

$$\psi^*(y) = \begin{cases} \sum_{i=1}^{n} -\log(-y_i) - 1 & \text{if } -y \in \mathbb{R}_{++}^n, \\ +\infty & \text{otherwise} \end{cases}$$

and, as already mentioned in Example 3, the distance is

$$D_\psi(x,y) = \sum_{i=1}^{n} -\log \frac{x_i}{y_i} + \frac{x_i}{y_i} - 1 \ \forall \ (x,y) \in \mathbb{R}_{++}^n \times \mathbb{R}_{++}^n.$$

4. For $\alpha \in (0,1)$, consider the family of functions

$$K_\alpha(t) = \begin{cases} (\alpha t - t^\alpha)/(1-\alpha) & \text{if } t \geq 0 \\ \infty & \text{otherwise.} \end{cases}$$

For $y \in \text{dom } \psi^* = (-\infty, \beta]^n$, where $\beta > 0$ and $\beta - \alpha = \alpha\beta$ one can easily compute $\psi_\beta^*(y) = \sum_{i=1}^{n}(1-y_i/\beta)^{-\beta}$. The corresponding distance is given by

$$D_\alpha(x,y) = \sum_{i=1}^{n} y_i^{\alpha-1}(y_i + \beta x_i) - (\beta+1)x_i^\alpha \ \text{on} \ \mathbb{R}_+^n \times \mathbb{R}_{++}^n.$$

The particular choice $\alpha = 1/2$, leads to the distance

$$D_{\frac{1}{2}}(x,y) = \sum_{i=1}^{n} \frac{(\sqrt{x_i} - \sqrt{y_i})^2}{\sqrt{y_i}}$$

5. For

$$K(t) = \begin{cases} t\log t - (1+t)\log(1+t) + (1+t)\log 2 & \text{if } t \geq 0 \\ \infty & \text{otherwise} \end{cases}$$

the conjugate ψ^* and its domain are given by

$$\psi^*(y) = -\sum_{i=1}^{n} \log(2 - e^{y_i}) \quad \text{and dom } \psi^* = (-\infty, \log 2)^n.$$

The corresponding distance is given by

$$D_\psi(x, y) = \sum_{i=1}^{n} x_i \log \frac{x_i}{y_i} - (1 + x_i) \log \left(\frac{1 + x_i}{1 + y_i} \right) \quad \text{on } \mathbb{R}_+^n \times \mathbb{R}_{++}^n.$$

Remark 1. It can be verified that the φ-divergence coincides with the separable Bregman distance, if and only if the corresponding kernel for both distances is the entropy kernel

$$\varphi(t) = t \log t - t + 1, \text{ i.e., with } \psi(x) = \sum_{i=1}^{n} \varphi(x_i).$$

In this case one has $D_\psi(x, y) = d_\varphi(x, y)$, and hence in both cases the resulting distance is the KL divergence.

We end this section by mentioning a simple but key property satisfied by the Bregman distance and revealed in [107]. This property plays a crucial role in the analysis of optimization problems (applications to clustering are outlined in Sect. 4, Example 9). This identity is called the *three point identity* as it appears to be a natural generalization of the quadratic identity valid for the Euclidean norm (i.e., of the law of cosines), and follows by direct substitution using (8).

Lemma 1. *[107, Lemma 3.1] For any three points $a, b \in int(\text{dom } \psi)$, and $c \in \text{dom } \psi$ the following identity holds true*

$$D_\psi(c, b) - D_\psi(c, a) = D_\psi(a, b) + \langle \nabla\psi(a) - \nabla\psi(b), c - a \rangle. \tag{17}$$

Unfortunately, such an identity does not hold for φ-divergences, yet some related useful inequalities do hold, see in particular [35, 417].
We now use the above framework to consider the k-means clustering type algorithms within this unified view point.

4 Entropic Means and k-Means Type Algorithms

In this section we state the batch k-means clustering algorithm with a general distance-like function d and derive formulas for centroids for various specific choices of d.

4.1 Batch k-Means Algorithm

We denote the standard simplex in $\mathbb{R}^k$ by Δ, i.e.,

$$\Delta = \left\{ w \in \mathbb{R}^k \ : \ w_j \geq 0, \ \sum_{j=1}^{k} w_j = 1 \right\}.$$

Given an entropy-like distance d, the batch k-means like clustering algorithm can be defined as follows.

Algorithm 1 (Batch k-means clustering algorithm)
Let $\{a^1, \ldots, a^m\}$ be the set of points in S to be clustered, and let $\{x^l(0) \in S : l = 1, \ldots, k\}$ be the k initial centroids.

Step 0 *Set $t = 0$.*
Step 1 *For $i = 1, \ldots, m$ solve*

$$w^i(t) = \underset{w \in \Delta}{\operatorname{argmin}} \sum_{l=1}^{k} w_l d(x^l(t), a^i).$$

Step 2 *Update the cluster centers by solving*

$$\left(x^1(t+1), \ldots, x^k(t+1) \right) = \underset{x^1, \ldots, x^k \in S}{\operatorname{argmin}} \left\{ \sum_{i=1}^{m} \sum_{l=1}^{k} w_l^i(t) d(x^l, a^i) \right\}.$$

Step 3 *If (stopping criterion fails) then*
increment t by 1, and go to Step 1 above.
Step 4 *Stop.*

Step 1 is trivially solved. For each $i = 1, \ldots, m$, let $l(i) = \underset{1 \leq l \leq k}{\operatorname{argmin}} d(x^l(t), a^i)$, breaking ties arbitrarily. The optimal $w^i(t)$ is simply given by $w^i_{l(i)}(t) = 1$ and $w^i_l(t) = 0, \forall l \neq l(i)$. It is not unusual to have $w^i_l(t) = 0$ for some $l \in \{1, \ldots, k\}$ and each $i = 1, \ldots, m$ (empty cluster l). In such a case the procedure can be restarted with a different initial partition. Another way to avoid empty clusters is to use the incremental version of the algorithm given in Sect. 5.2.

Example 5. (The quadratic case). Suppose that $S = \mathbb{R}^n$ and $d(x, y) = \frac{1}{2} \| x - y \|^2$. Then, Step 2 is solved by equating to zero the gradient of the objective function with respect to $x^1, \ldots, x^k$. This yields immediately the *arithmetic mean* formula

$$x^l(t+1) = \frac{\displaystyle\sum_{i=1}^{m} w^i_{l(i)}(t) a^i}{\displaystyle\sum_{i=1}^{m} w^i_{l(i)}(t)}.$$

From this, it is clear that the mechanism of the k-means algorithm can be applied the same way to problems with constraints $S = \mathbb{R}^n_+$ by using the entropy-like distance functions given in Sect. 3. Indeed, Step 1 remains unchanged while Step 2 (with d smooth enough that satisfies the gradient properties, i.e., with $d(\cdot, a)$ essentially smooth, the constraints are automatically eliminated, and the minimizer belongs to $\mathbb{R}^n_{++}$) requires to solve for x^l an equation of the form

$$\sum_{i=1}^{m} w^i_{l(i)}(t)\nabla_{x^l} d(x^l, a^i) = 0.$$

It turns out that for the distance-like functions proposed in Sect. 3, this equation can also be solved analytically for various choices of the kernels φ, ψ. In fact due to the structure of the equation given here, and the separability involved in the examples of entropy-like distances, it is enough to consider the scalar case, namely to solve a scalar equation of the form (with some abuse of notations):

$$\sum_{i=1}^{m} w_i d'(x, a_i) = 0, \quad \text{(where } d'(x, a_i) \text{ denotes the derivative with respect to } x).$$

This reduces the centroid computations to the so-called entropic means of Ben-Tal et al. [56], which we now briefly recall.

4.2 Characterization of Means

Ben-Tal et al. [56] have shown how to generate means as optimal solutions of a minimization problem with a "distance" function as objective. Let $a = (a_1, \ldots, a_m)$ be given strictly positive numbers and let $w = (w_1, \ldots, w_m)$ be given weights, i.e., $\sum w_i = 1$, $w_i > 0, i = 1, \ldots, n$.

The mean of $(a_1, \ldots, a_n)$ is defined as the value of x for which the sum of distances from x to each a_i (denoted "dist"(x, a_i)) is minimized, i.e., the mean is the optimal solution of

$$(E) \quad \min\left\{\sum w_i \text{"dist"}(x, a_i) : x \in \mathbb{R}_+\right\}.$$

The notation "dist"(α, β) refers here to some measure of distance between $\alpha, \beta > 0$, which must satisfy

$$\text{"dist"}(\alpha, \beta) = 0, \text{ if } \alpha = \beta \quad \text{and} \quad \text{"dist"}(\alpha, \beta) > 0, \text{ if } \alpha \neq \beta.$$

Adopting the φ-divergence we define the distance from x to a_i by

$$\text{"dist"}(x, a_i) := d_\varphi(x, a_i) = a_i \varphi(x/a_i) \tag{18}$$

for each $i = 1, \ldots, n$. The optimization problem (E) is now

$$(E_{d_\varphi}) \quad \min \left\{ \sum_{i=1}^{m} w_i a_i \varphi\left(\frac{x}{a_i}\right) : x \in \mathbb{R}_+ \right\},$$

and the resulting optimal solution denoted $\bar{x}_\varphi(a) := \bar{x}_\varphi(a_1, \ldots, a_m)$ is called the *entropic mean* of $(a_1, \ldots, a_n)$, and is obtained by solving:

$$\sum_{i=1}^{m} w_i \varphi'\left(\frac{x}{a_i}\right) = 0. \tag{19}$$

It is shown in [56] that the entropic mean satisfies the basic properties of a general mean and that all classical means as well as many others are special cases of entropic means.

Example 6. The choice of $\varphi(t) = 1/(p-1)(t^{1-p} - p) + t$ in (E) leads to the so-called mean of order p,

$$\bar{x}_p(a) = \left(\sum_{i=1}^{m} w_i a_i^p\right)^{1/p}.$$

The means of order $1, \frac{1}{2}, 0$, which are the arithmetic mean, the root mean square, and the geometric mean can be characterized by choosing the kernels

$$-\log t + t - 1, \ 1 - 2\sqrt{t} + t, \ \text{and} \ t \log t - t + 1,$$

respectively. Less popular means like the Lehmer mean

$$\bar{x}_L(a) = \frac{\displaystyle\sum_{i=1}^{m} w_i a_i^p}{\displaystyle\sum_{i=1}^{m} w_i a_i^{p-1}}$$

and the Gini mean:

$$\bar{x}_G(a) = \left(\frac{\sum w_i a_i^s}{\sum w_i a_i^r}\right)^{\frac{1}{s-r}} \quad \text{for } s \geq 0 > r \text{ or } s > 0 \geq r.$$

can be obtained by solving problem (E) with

$$\varphi_L(t) = \frac{t^{2-p}}{2-p} - \frac{t^{1-p}}{1-p} + \frac{1}{(2-p)(1-p)}$$

and

$$\varphi_G(t) = \frac{t^{1-r} - 1}{1-r} - \frac{t^{1-s} - 1}{1-s},$$

respectively. These results can also be extended to derive entropic means for random variables, see [56] for more details.

Likewise, adopting the Bregman distance

$$\text{``dist''}(\alpha, \beta) = D_\psi(\alpha, \beta) := \psi(\alpha) - \psi(\beta) - (\alpha - \beta)\psi'(\beta), \qquad (20)$$

one has to solve

$$(E_{D_\psi}) \qquad \min\left\{\sum_{i=1}^{m} w_i D_\psi(x, a_i) : x \in \text{dom } \psi\right\}.$$

Substituting D_ψ in the objective function of (E_{D_ψ}) we have to solve (since the a_i are given numbers) the *convex* minimization problem:

$$\min\left\{\psi(x) - x\sum_{i=1}^{m} w_i\psi'(a_i) : x \in \text{dom } \psi\right\}.$$

Under the set-up of Sect. 3.2 (i.e., recall that $\psi'(t) \to -\infty$ as $t \to \bar{t}$ a boundary point of dom ψ), one has $\bar{x}_\psi := \bar{x}_\psi(a)$ solves (E_{D_ψ}) if and only if

$$\psi'(\bar{x}_\psi) = \sum_{i=1}^{m} w_i\psi'(a_i). \qquad (21)$$

Since ψ' is continuous and strictly increasing on int(dom ψ), it follows that there is a unique $\bar{x}_\psi$ solving (21) given by

$$\bar{x}_\psi(a) = (\psi')^{-1}\left\{\sum_{i=1}^{m} w_i\psi'(a_i)\right\}, \qquad (22)$$

where $(\psi')^{-1}$ denotes the inverse function of ψ', which exits under our assumptions on ψ (cf. Sect. 3.2). Hence with $\psi' := h$, h is strictly monotone and we have characterized the so-called generalized mean of Hardy, Littlewood, and Polya (HLP) [220]:

$$\bar{x}_h(a) = h^{-1}\left\{\sum_{i=1}^{m} w_i h(a_i)\right\}. \qquad (23)$$

For more details, we refer the reader to [56].

4.3 Centroid Computation: Examples

Given the separable structure of the divergences (see (15) and Example 4), computations of centroids in the multidimensional case follow straightforward from results presented earlier. For the φ-divergences we have given a number of interesting examples that can be adapted to produce the corresponding centroids.

For the case of Bregman distances, the formula is explicitly given by (22), and can be used right away as long as we have a formula for $(\psi')^{-1} = (\psi^*)'$ (see (9)), which is the case for all the kernels given in Example 4.

We complete this section by mentioning three more examples of particular interest.

Example 7. First, consider the second-order divergence (see (6)) that can also be used in problem (E). In this case the optimality condition

$$\nabla_x \left(\sum_{i=1}^m w_i a_i^2 \varphi(x/a_i) \right) = 0$$

yields

$$\sum_{i=1}^m w_i a_i \varphi'(x/a_i) = 0.$$

Using the kernel

$$\varphi_4(t) = \frac{\nu}{2}(t-1)^2 + \mu(-\log t + t - 1),$$

which was introduced in Sect. 3.1 we obtain the unusual mean:

$$x_{\mu,\nu} = \frac{(\nu-\mu)E(A) + \sqrt{(\nu-\mu)^2 E^2(A) + 4\mu\nu E(A^2)}}{2\nu}, \tag{24}$$

where $E(A) := \sum_{i=1}^m w_i a_i$, $E(A^2) := \sum_{i=1}^m w_i a_i^2$. The quantities $E(A)$ and $E(A^2)$ can be interpreted as the first (mean) and second moments of the random variable A with $\text{Prob}\{A = a_i\} = w_i$. The limiting case $\mu \to 0$ recovers the standard weighted arithmetic mean used in the k-means algorithm with the squared Euclidean distance.

Example 8. Now consider a *regularized ϕ-divergence distance*, namely a distance of the form

$$d_R(x,y) := \frac{\nu}{2}\|x-y\|^2 + d_\varphi(x,y)$$

with $\varphi(t) = \varphi_2(t) = -\log t + t - 1$ (note that this is no more a second-order homogeneous distance). A simple computation shows that the resulting mean is just the weighted arithmetic mean. This of course should not be too surprising, since the above is a positive linear combination of two distances, each of which has been shown previously to yield the arithmetic mean. Indeed, recall from Sect. 3.1 that for φ_2, the resulting distance is just $KL(y,x)$ (i.e., KL divergence with reversed order of variables). The joint convexity of the φ-divergence, allows to reverse the order of the variables, or likewise to use a kernel $\hat{\varphi}(t) := t\varphi(1/t)$, e.g., with $\varphi(t) = t\log t - t + 1$, one has $\hat{\varphi}(t) = \varphi_2(t)$. Alternatively, with $\psi(t) = \nu t^2/2 + t\log t - t + 1$, one can write $d_R(x,y) = D_\psi(y,x)$.

This example raises the natural question: "what happens with Bregman distances when the order of the variables is reversed?" Indeed, since a Bregman distance is not necessarily symmetric we can get a different centroid when minimizing with respect to the second argument. This approach has been recently adopted in [46, 47] who have provided the following interesting and somewhat surprising answer.

Example 9. When reversing the order of the variables in a Bregman distance, the resulting function to be minimized is, in general, not necessarily convex in the second argument (except for the trivial choices $\psi(t) = t^2/2$ and $\psi(t) = t \log t - t$ or/and by imposing further conditions on ψ, which unfortunately preclude the use of the interesting examples), i.e., one has to solve in general the nonconvex problem

$$\min \left\{ \sum_{i=1}^{m} w_i D_\psi(a_i, x) : x \in \mathrm{int}(\mathrm{dom}\ \psi) \right\}. \tag{25}$$

However, even though the problem (25) is nonconvex, it has been shown by Banerjee et al. [46] that the global minimizer of (25) is *always* the standard weighted arithmetic mean. This can be seen as follows. Due to Lemma 1, and the fact that $D_\psi(\cdot, \cdot) \geq 0$, one has

$$D_\psi(a_i, x) - D_\psi(a_i, z) = D_\psi(z, x) + (\psi'(z) - \psi'(x))(a_i - z)$$
$$\geq (\psi'(z) - \psi'(x))(a_i - z).$$

Thus, multiplying the above by w_i and summing the resulting inequalities it immediately follows that with $z := \sum_{i=1}^{m} w_i a_i$, the right-hand side of the inequality is 0 and

$$\sum_{i=1}^{m} w_i D_\psi(a_i, x) \geq \sum_{i=1}^{m} w_i D_\psi \left(a_i, \sum_{i=1}^{m} w_i a_i \right) \qquad \forall x \in \mathrm{int}(\mathrm{dom}\ \psi).$$

Hence $z := \sum_{i=1}^{m} w_i a_i$ is the global minimizer, and for the Bregman distance with the reversed order of variables we always obtain the standard weighted arithmetic mean for the corresponding centroid. Further interesting results when minimizing a Bregman distance with respect to the second argument have been developed in [47]. For example, an explicit connection of Bregman distances to the exponential family of statistical distributions has also been shown there.

We next consider a special particular case of normalized KL distance.

Example 10. Consider the KL distance (see Example 2) obtained from (13) with $\nu = 0$, $\mu = 1$, i.e.,

$$D_\psi(y, x) = \sum_{j=1}^{n} y_j \log \frac{y_j}{x_j} + \sum_{j=1}^{n} x_j - \sum_{j=1}^{n} y_j$$

(see (14)). When the vectors with non-negative coordinates x, y are L_1 normalized $D_\psi(y, x)$ reduces to

$$KL(y, x) = \sum_{j=1}^{n} y_j \log \frac{y_j}{x_j}.$$

Hence if a^i are L_1 norm vectors with non-negative coordinates,

1. $M_1 = \min \left\{ \sum_{i=1}^{m} w_i D_\psi(a^i, x) \ : \ 0 \leq x_j \right\}$

   and

2. $M_2 = \min \left\{ \sum_{i=1}^{m} w_i KL(a^i, x) \ : \ 0 \leq x_j, \ \sum_{j=1}^{n} x_j = 1 \right\}$

   $= \min \left\{ \sum_{i=1}^{m} w_i D_\psi(a^i, x) \ : \ 0 \leq x_j, \ \sum_{j=1}^{n} x_j = 1 \right\},$

then one has $M_1 \leq M_2$. Since as shown in Example 9

$$M_1 = \sum_{i=1}^{m} w_i D_\psi(a^i, z) \text{ where } z = \sum_{i=1}^{m} w_i a^i,$$

one has $0 \leq z_j$, $\sum_{j=1}^{n} z_j = 1$. This shows that $M_1 = M_2$, and the centroid for the case of the "simple" KL distance with L_1 normalized data is given by the arithmetic mean.

The above development justifies examining and analyzing the effect, benefits and usefulness of using the resulting family of k-means type algorithms equipped with entropy-like distances.

5 Batch and Incremental k-Means Algorithm

We now switch gears and state k-means clustering algorithms focusing on *partitions*, rather that on centroids. While the two approaches are equivalent (see Sect. 2), the "partition" view of the algorithm makes and some drawbacks of the batch k-means clustering algorithm apparent, and justifies introduction of the incremental k-means clustering algorithm and the merger of the algorithms (see Sect. 5.2).

5.1 Batch Algorithm

The batch k-means algorithm is the two-step procedure oscillating between:

1. computation of centroids $\{x^1, \ldots, x^k\}$ for a given partition Π, and
2. computation of a partition $\Pi = \{\pi_1, \ldots, \pi_k\}$ for a given set of centroids $\{x^1, \ldots, x^k\}$

as described in Sect. 2. An application of these two steps to a partition Π generates partition $\texttt{nextBKM}(\Pi)$ and $Q(\Pi) \geq Q(\texttt{nextBKM}(\Pi))$ (see [166] for the case of $d(x, y) = \|x - y\|^2$). While fast and easy to implement, the batch k-means algorithm often gets trapped at a local minimum. In Example 11, we consider a simple scalar dataset $\mathcal{A}$ and the batch k-means algorithm with the squared Euclidean distance.

Example 11. Let $\mathcal{A} = \{0, 2, 3\}$, and the initial partition $\Pi^{(0)} = \{\pi_1^{(0)}, \pi_2^{(0)}\}$ where $\pi_1^{(0)} = \{0, 2\}$, and $\pi_2^{(0)} = \{3\}$.

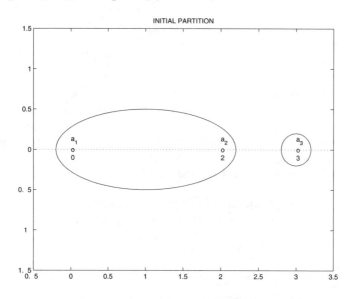

An iteration of the batch k-means algorithm applied to $\Pi^{(0)}$ does not change the partition. On the other hand the partition $\Pi^{(1)} = \{\{0\}, \{2, 3\}\}$ is superior to $\Pi^{(0)}$. The better partition $\Pi^{(1)}$ is undetected by the algorithm.

Next we continue to deal with $d(x, y) = \|x - y\|^2$ and describe a modification of the batch k-means algorithm introduced independently by [219, 277, 458].

5.2 Incremental Algorithm

An alternative form of the batch k-means algorithm, the incremental algorithm (see, e.g., [144]), is capable of "fixing" Example 11. For a given partition

$\Pi = \{\pi_1, \ldots, \pi_k\}$ of the dataset $\mathcal{A} = \{a^1, \ldots, a^m\}$ the algorithm examines partitions generated from Π by removing a single vector a from a cluster π_i and assigning this vector to a cluster π_j. We shall denote the "new" clusters by $\pi_i^- = \pi_i - \{a\}$ and $\pi_j^+ = \pi_j \cup \{a\}$, and the obtained partition by Π'. The vector a and the "destination" cluster π_j are associated with the change of the objective function Q, i.e.,

$$q(a, j) = Q(\Pi) - Q(\Pi') = [Q(\pi_i) - Q(\pi_i^-)] + [Q(\pi_j) - Q(\pi_j^+)]. \quad (26)$$

The partition Π' that maximizes (26) is called "first variation" of Π and denoted by $\texttt{nextFV}(\Pi)$. The main computational challenge associated with (26) is evaluation of $Q(\pi^+)$ and $Q(\pi^-)$. A convenient expression for (26) is given in [144]. Specifically,

Theorem 1. *Let* $\pi_i^- = \pi_i^- \cup \{a\}$ *and* $\pi_j^+ = \pi_j \cup \{a\}$. *If* $|\pi_i| = m_i$, $|\pi_j| = m_j$, $x^i = x(\pi_i)$, *and* $x^j = x(\pi_j)$, *then*

$$[Q(\pi_i) - Q(\pi_i^-)] + [Q(\pi_j) - Q(\pi_j^+)] = \frac{m_i}{m_i - 1} \left|\left| x^i - a \right|\right|^2$$
$$- \frac{m_j}{m_j + 1} \left|\left| x^j - a \right|\right|^2. \quad (27)$$

Expression (27) shows that to compute the change of the objective function due to an iteration of the incremental step one needs to know the cluster size and the distances from each point of the dataset to all the centroids. These distances are also computed at each iteration of the batch k-means algorithm. This observation suggests applying a series of batch k-means iterations followed by a single incremental iteration as follows:

Algorithm 2 (The k-means clustering algorithm)
For a user supplied non-negative tolerances $\texttt{tol}_B$ *and* $\texttt{tol}_I$ *do the following:*

1. *Start with an arbitrary partitioning* $\Pi^{(0)} = \left\{ \pi_1^{(0)}, \ldots, \pi_k^{(0)} \right\}$. *Set the index of iteration* $t = 0$.
2. *Generate the partition* $\texttt{nextBKM}\left(\Pi^{(t)}\right)$.
   if $\left[Q\left(\Pi^{(t)}\right) - Q\left(\texttt{nextBKM}\left(\Pi^{(t)}\right)\right) > \texttt{tol}_B \right]$
       set $\Pi^{(t+1)} = \texttt{nextBKM}\left(\Pi^{(t)}\right)$
       increment t *by 1.*
       go to 2
3. *Generate the partition* $\texttt{nextFV}\left(\Pi^{(t)}\right)$.
   if $\left[Q\left(\Pi^{(t)}\right) - Q\left(\texttt{nextFV}\left(\Pi^{(t)}\right)\right) > \texttt{tol}_I \right]$
       set $\Pi^{(t+1)} = \texttt{nextFV}\left(\Pi^{(t)}\right)$.
       increment t *by 1.*
       go to 2
4. *Stop.*

As a rule an iteration of the batch algorithm changes cluster affiliation for a large set of vectors, and an iteration of the incremental algorithm moves no more than one vector. The above described "merger" of the batch and incremental versions of the k-means algorithm enjoys speed of the batch version and accuracy of the incremental version (unlike the incremental algorithm introduced in [144], which looks for any vector whose reassignment leads to an improvement of the objective function we employ – the "greedy" version that looks for the maximal possible drop of the objective function). We next apply Algorithm 2 to the initial partition provided in Example 11.

Example 12. A single iteration of the first variation applied to the initial partition $\Pi^{(0)} = \{\{0,2\}, \{3\}\}$ generates the optimal partition $\Pi^{(1)} = \{\{0\}, \{2,3\}\}$.

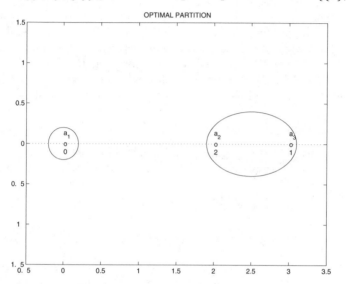

Note that all numerical computations associated with Step 3 of Algorithm 2 have been already performed at Step 2 (see (27)). The improvement over batch k-means comes, therefore, at virtually no additional computational expense.

The decision whether a vector $a \in \pi_i$ should be moved from cluster π_i with m_i vectors to cluster π_j with m_j vectors is made by the batch k-means algorithm based on examination of the expression $||a - x^i|| - ||a - x^j||$. The positive sign of

$$\Delta_k = ||a - x^i||^2 - ||a - x^j||^2,$$

may trigger the move. As (27) shows, the change in the value of the objective function caused by the move is

$$\Delta = \frac{m_i}{m_i - 1} ||a - x^i||^2 - \frac{m_j}{m_j + 1} ||a - x^j||^2.$$

The difference between the expressions

$$\Delta - \Delta_k = \frac{1}{m_i - 1} \left\| a - x^i \right\|^2 + \frac{1}{m_j + 1} \left\| a - x^j \right\|^2 \geq 0 \qquad (28)$$

is negligible when m_i and m_j are large numbers. However $\Delta - \Delta_k$ may become significant for small clusters. In particular, it is possible that Δ_k is negative, and the batch k-means iteration leaves a in cluster π_i. At the same time the value of Δ is positive, and reassigning a to π_j would decrease Q. Indeed, for the dataset of Example 11 and $a = 2$ one has

$$\Delta_k = \|a^2 - x^1\|^2 - \|a^2 - x^2\|^2 = 1 - 1 = 0,$$

and

$$\Delta = \frac{2}{1} \|a^2 - x^1\|^2 - \frac{1}{2} \|a^2 - x^2\|^2 = 2 - \frac{1}{2} = \frac{3}{2}.$$

Note that the discussion following Example 12 holds true for a general distance-like function $d(x, y)$ (a detailed analysis of incremental k-means with KL divergence is offered in [58]). Indeed, given a partition Π the decision whether a vector $a \in \pi_i$ should be moved from cluster π_i to cluster π_j is made by the batch k-means algorithm based on examination of the expression

$$\Delta_k = d(x^i, a) - d(x^j, a). \qquad (29)$$

We denote centroids $x(\pi_i^-)$ and $x(\pi_j^+)$ by $(x^i)^-$ and $(x^j)^+$, respectively. The exact change in the value of the objective function caused by the move is

$$\Delta = \left[Q(\pi_i) - Q(\pi_i^-) \right] + \left[Q(\pi_j) - Q(\pi_j^+) \right]$$

$$= \left[\sum_{a' \in \pi_i^-} d(x^i, a') - \sum_{a' \in \pi_i^-} d((x^i)^-, a') \right] + d(x^i, a)$$

$$+ \left[\sum_{a' \in \pi_j^+} d(x^j, a') - \sum_{a' \in \pi_j^+} d((x^j)^+, a') \right] - d(x^j, a).$$

Hence $\Delta - \Delta_k$ is given by

$$\sum_{a' \in \pi_i^-} d(x^i, a) - \sum_{a' \in \pi_i^-} d((x^i)^-, a) + \sum_{a' \in \pi_j^+} d(x^j, a) - \sum_{a' \in \pi_j^+} d((x^j)^+, a'). \quad (30)$$

Due to definition of $x(\pi)$ both differences in (30) are non-negative, hence

$$\Delta - \Delta_k \geq 0. \qquad (31)$$

If $d(x, a)$ is strictly convex in x, then the inequality in (31) is strict provided at least one new centroid is different from the corresponding old centroid. This observation justifies application of an incremental iteration following a series of batch iterations for general distance-like functions $d(x, y)$ as suggested in Algorithm 2 for $d(x,y) = \|x - y\|^2$. Batch k-means with KL divergence was already introduced in [137], its incremental counterpart appears in [58], and the batch algorithm augmented by the incremental step was suggested in [130, 279].

Note that removal/assignment of a vector from/to a cluster changes clusters' centroids and all distances from the vectors in the two clusters involved to the "new" centroids. Often the actual computations needed to evaluate the change in the objective function use the distances between the "old" centroid and the vector changing its cluster affiliation only.

Example 13. Incremental step with Bregman distance with reversed order of arguments, i.e., $d(x,y) = D_\psi(y,x)$.

Let π be a set with p vectors, and $a \in \pi$. We denote the set $\pi - \{a\}$ by π^-, and centroids of the sets π and π^- (which are given by the arithmetic mean) by x and x^-, respectively. We first provide a formula for $Q(\pi) - Q(\pi^-)$. Note that

$$
\begin{aligned}
Q(\pi) - Q(\pi^-) &= \sum_{a' \in \pi^-} d(x, a') + d(x, a) - \sum_{a' \in \pi^-} d(x^-, a') \\
&= \sum_{a' \in \pi^-} [\psi(a') - \psi(x) - \langle a' - x, \nabla\psi(x)\rangle] \\
&\quad - \sum_{a' \in \pi^-} \left[\psi(a') - \psi(x^-) - \langle a' - x^-, \nabla\psi(x^-)\rangle\right] + d(x, a) \\
&= (p - 1)\left[\psi(x^-) - \psi(x)\right] - \langle \sum_{a' \in \pi^-} a' - (p-1)x, \nabla\psi(x)\rangle \\
&\quad + d(x, a).
\end{aligned}
$$

Keeping in mind that

$$
\sum_{a' \in \pi^-} a' = \sum_{a' \in \pi} a' - a = px - a,
$$

we get

$$
Q(\pi) - Q(\pi^-) = (p - 1)\left[\psi(x^-) - \psi(x)\right] - \langle x - a, \nabla\psi(x)\rangle + d(x, a). \quad (32)
$$

Next we derive a convenient formula for $Q(\pi) - Q(\pi^+)$, where π is a set with p vectors, $a \notin \pi$, $\pi^+ = \pi \cup \{a\}$, $x = x(\pi)$, and $x^+ = x(\pi^+)$.

$$
\begin{aligned}
Q(\pi) - Q(\pi^+) &= \sum_{a' \in \pi^+} d(x, a') - d(x, a) - \sum_{a' \in \pi^+} d(x^+, a') \\
&= \sum_{a' \in \pi^+} [\psi(a') - \psi(x) - \langle a' - x, \nabla\psi(x)\rangle] \\
&\quad - \sum_{a' \in \pi^+} \left[\psi(a') - \psi(x^+) - \langle a' - x^+, \nabla\psi(x^+)\rangle\right] - d(x, a)
\end{aligned}
$$

$$= (p+1)\left[\psi(x^+) - \psi(x)\right] - \left\langle \sum_{a' \in \pi^+} a' - (p+1)x, \nabla\psi(x)\right\rangle$$

$$-d(x,a).$$

Keeping in mind that

$$\sum_{a' \in \pi^+} a' = \sum_{a' \in \pi} a' + a = px + a,$$

we obtain

$$Q(\pi) - Q(\pi^+) = (p+1)\left[\psi(x^+) - \psi(x)\right] - \langle a - x, \nabla\psi(x)\rangle - d(x,a). \quad (33)$$

Finally we select two clusters π_i, and π_j from a partition Π and $a \in \pi_i$. We denote the number of vectors in each cluster by m_i and m_j, respectively. The formula for

$$\left[Q(\pi_i) - Q(\pi_i^-)\right] + \left[Q(\pi_j) - Q(\pi_j^+)\right]$$

follows from (32) and (33) and is given by

$$
\begin{aligned}
\left[Q(\pi_i) - Q(\pi_i^-)\right] + \left[Q(\pi_j) - Q(\pi_j^+)\right] &= (m_i - 1)\left[\psi((x^i)^-) - \psi(x^i)\right] \\
&\quad -\langle x^i - a, \nabla\psi(x^i)\rangle \\
&\quad +(m_j + 1)\left[\psi((x^j)^+) - \psi(x^j)\right] \\
&\quad +\langle x^j - a, \nabla\psi(x^j)\rangle. \quad (34)
\end{aligned}
$$

In text mining applications, due to sparsity of the data vector a, most co-ordinates of centroids $(x^i)^-$ and x^i coincide. Hence, when the function ψ is separable, computations of $\psi((x^i)^-) - \psi(x^i)$ are relatively cheap. On the other hand, the quantities $\langle x^i - a, \nabla\psi(x^i)\rangle$ have already been computed at the batch step of the k-means algorithm and can be used by the incremental step at no additional computational cost. (This remark holds true also for centroids $(x^j)^+$ and x^j).

While the computational cost of an incremental iteration following a batch iteration remains negligible for distance-like functions generated by Bregman distances with reversed order of variables, it may be as expensive as the cost of a batch iteration for distance-like functions generated by Csiszar divergences (see, e.g., [280]).

5.3 BIRCH-Like Extensions with Entropy-Like Distances

When the dataset $\mathcal{A} = \{a^1, \ldots, a^m\}$ is very large, one can consider the following clustering strategy:

1. Partition the dataset into a large number of small dense clusters,
2. Treat each small cluster as a single "cluster-point" and cluster the set of "cluster-points," and

3. Recover the partition of the original datapoints from the generated clusters of "cluster-points."

The suggested strategy of sequential application of a number of clustering algorithms was implemented in [276, 278]. The first step of the described procedure can be implemented using, for example, the BIRCH strategy through adaptive construction of a balanced cluster features tree in a preassigned memory buffer (see [459]). In what follows we use PDDP [76] to simulate data-squashing capability of BIRCH. If, for example, the average size of obtained small clusters is 10, the size of the dataset is reduced by an order of magnitude. The main technical questions one has to address are:

1. What information is needed for clustering "cluster-points,"
2. How much savings (in terms of speed and memory) can be achieved using this approach.

The answers to these two questions for $d(x, y) = \|x - y\|^2$ are provided in [459] and are summarized in the following result that follows from [144] (see also Theorem 1).

Theorem 2. If $\mathcal{A} = \pi_1 \cup \pi_2 \cup \ldots \cup \pi_k$ with $m_i = |\pi_i|$, $\pi_i \cap \pi_j = \emptyset$ when $i \neq j$, and $x^i = x(\pi_i)$, $i = 1, \ldots, k$, then

$$x = x(\mathcal{A}) = \frac{m_1}{m} x^1 + \ldots + \frac{m_k}{m} x^k, \text{ where } m = m_1 + \ldots + m_k$$

and

$$Q(\mathcal{A}) = \sum_{i=1}^{k} Q(\pi_i) + \sum_{i=1}^{k} m_i \left\| x - x^i \right\|^2. \tag{35}$$

If

$$\mathcal{X} = \{\underbrace{x^1, \ldots, x^1}_{m_1}, \ldots, \underbrace{x^k, \ldots, x^k}_{m_k}\}$$

is the set of centroids counted with appropriate weights, then (35) leads to the following

$$Q(\mathcal{A}) - Q(\mathcal{X}) = \sum_{i=1}^{k} Q(\pi_i).$$

The theorem shows that the quality of a partition comprising clusters $\pi_1, \ldots, \pi_k$ is a function of:

1. Cluster centroids x^i, $i = 1, \ldots, k$;
2. Cluster sizes $m_i = |\pi_i|$, $i = 1, \ldots, k$; and
3. Cluster qualities $Q(\pi_i)$, $i = 1, \ldots, k$.

Hence the required overhead information per cluster is just two scalars $|\pi_i|$ and $Q(\pi_i)$, and so m_i, x^i, and $Q(\pi_i)$ are sufficient statistics.

Next we consider a couple of examples where the BIRCH construction can be applied to distances other than $\|x-y\|^2$. First note that if $\pi = \{a^1, \ldots, a^p\}$, $x(\pi)$ is the centroid, and x is a vector, then

$$\sum_{i=1}^{p} d(x, a^i) = \sum_{i=1}^{p} d(x(\pi), a^i) + \left[\sum_{i=1}^{p} d(x, a^i) - \sum_{i=1}^{p} d(x(\pi), a^i)\right]$$

$$= Q(\pi) + \left[\sum_{i=1}^{p} d(x, a^i) - \sum_{i=1}^{p} d(x(\pi), a^i)\right]. \tag{36}$$

The first example handles Bregman distances with reversed order of variables.

Example 14. Consider the distance-like function $d(x, a) = D_\psi(a, x) = \psi(a) - \psi(x) - \langle a - x, \nabla\psi(x)\rangle$ (see Example 9). Due to (36), one has

$$\sum_{i=1}^{p} d(x, a^i) = Q(\pi) + p\left[\psi(x(\pi)) - \psi(x) - \langle x(\pi) - x, \nabla\psi(x)\rangle\right]$$

$$= Q(\pi) + pD_\psi(x(\pi), x).$$

This straightforward computation immediately leads to the following generalization of Theorem 2.

Theorem 3. *If* $\mathcal{A} = \pi_1 \cup \pi_2 \cup \cdots \cup \pi_k$ *with* $m_i = |\pi_i|$, $x^i = x(\pi_i)$, $i = 1, \ldots, k$; *and*

$$x = x(\mathcal{A}) = \frac{m_1}{m} x^1 + \cdots + \frac{m_k}{m} x^k, \ \text{where } m = m_1 + \cdots + m_k,$$

then

$$Q(\mathcal{A}) = \sum_{i=1}^{k} Q(\pi_i) + \sum_{i=1}^{k} m_i \left[\psi(x^i) - \psi(x)\right]$$

$$= \sum_{i=1}^{k} Q(\pi_i) + \sum_{i=1}^{k} m_i D_\psi(x(\pi_i), x^i).$$

The required overhead information per cluster are two scalars $|\pi_i|$ and $Q(\pi_i)$.

We now turn to a distance-like function generated by Csiszar divergence.

Example 15. Let $\varphi_1(t) = t \log t - t + 1$. The corresponding distance-like function is (see (4))

$$d(x, a) = \sum_{j=1}^{n} \left[x_j \log \frac{x_j}{a_j} - x_j + a_j\right].$$

The centroid $x = x(\pi)$ of a p element set π is given by the geometric mean, i.e.,

$$x_j = \begin{cases} \prod_{a \in \pi} (a_j)^{\frac{1}{p}} & \text{if } a_j > 0 \ \forall \ a \in \pi, \\[2ex] 0 & \text{otherwise} \end{cases}$$

(for details and clustering with geometric means see [280]).

Given a set of p vectors $\mathcal{A} = \{a^1, \ldots, a^p\}$ and a set of q vectors $\mathcal{B} = \{b^1, \ldots, b^q\}$, denote $x(\mathcal{A})$ by x^1, $x(\mathcal{B})$ by x^2, and $x(\mathcal{A} \cup \mathcal{B})$ by x. Note that

$$x_j = \begin{cases} 0 & \text{if} \qquad x_j^1 x_j^2 = 0, \\[2ex] \left[(x_j^1)^p (x_j^2)^q \right]^{\frac{1}{p+q}} & \text{otherwise.} \end{cases}$$

A straightforward computation shows that the quality of the cluster $\mathcal{A} \cup \mathcal{B}$ is

$$Q(\mathcal{A} \cup \mathcal{B}) = \sum_{i=1}^{p} e^T a^i + \sum_{i=1}^{q} e^T b^i - (p+q) e^T x,$$

where e is the vector of ones. The formula shows that in order to employ the BIRCH type clustering procedure one needs to have the following information concerning each cluster π:

1. Centroid $x(\pi)$,
2. Cluster size $|\pi|$, and
3. $e^T \left(\sum_{a \in \pi} a \right)$.

Note that the required overhead information per cluster is two scalars, exactly the same as in the quadratic case.

6 Experimental Results

In this section we present numerical experiments to show the effectiveness of the merger of batch and incremental k-means algorithms on gene microarray data and the efficiency of our BIRCH-like approach on text data. All the experiments are performed on a Linux machine with 2.4 GHz Pentium IV processor and 1 GB main memory.

Recent microarray technology has made it possible to monitor the expression levels of tens of thousands of genes in parallel. The outcome of a microarray experiment is often represented as a matrix called *gene-expression matrix*. Within a gene-expression matrix, rows correspond to genes and columns often represent samples related to some diseases such as cancers. Each entry gives the expression level of a gene in a sample. Clustering of cancer-related samples according to their expression across all or selected genes can characterize expression patterns in normal or diseased cells and thus help improve cancer treatment [199]. In our experiments, we use two human cancer microarray data sets: *lung cancer data*[2] and *leukemia data*[3]. The lung cancer data consist of 31 malignant pleural mesothelioma (MPM) samples and 150 adenocarcinoma (ADCA) samples [201]. Each sample is described by 12,533 genes. The leukemia data consists of 72 samples, including 47 acute lymphoblastic leukemia (ALL) samples and 25 acute myeloid leukemia (AML) samples [199]. Each sample is described by 6,817 genes. For both data sets, we perform a preprocessing to remove the genes that exhibit near-constant expression levels across all the samples, based on the ratio and the difference of each gene's maximum and minimum expression values, denoted as max/min and $max - min$. For lung cancer, after filtering out genes with $max/min < 5$ and $max - min < 500$, we are left with a $2,401 \times 181$ gene-expression matrix. For the leukemia data, we remove genes with $max/min < 5$ and $max - min < 600$ and are left with a $3,571 \times 72$ gene-expression matrix. Then we normalize each gene (row) vector in both matrices such that each row has mean 0 and standard derivation 1, i.e., if g^i is the original gene vector, then the normalized gene vector $\hat{g}^i$ is given by

$$\hat{g}^i_j = \frac{g^i_j - \bar{g}_i}{\sigma_i},$$

where $\bar{g}_i = \sum_{j=1}^s g^i_j / s$, $\sigma_i = \sqrt{\sum_{j=1}^s (g^i_j - \bar{g}_i)^2 / s}$, and s is the number of samples.

To evaluate the clustering of samples, we compare it with the sample class labels that we already know and form a *confusion matrix*, in which entry (i, j), $n_i^{(j)}$, gives the number of samples in cluster i that belong to class j. A diagonal confusion matrix is desirable since it represents a perfect clustering.

We apply the batch Euclidean k-means, i.e., k-means with squared Euclidean distances (C_2), on the lung cancer data, with random initialization, 100 times. On average, 45 samples are misclassified according to the class labels. However, if we apply the merger of the batch and the incremental Euclidean k-means algorithm described in Algorithm 1 on lung cancer data, we get perfect clustering for all 100 random runs. Figure 1 shows the objective function values with the progress of batch and incremental iterations on a particular run of Algorithm 1. A circle indicates a batch iteration and a star indicates an incremental iteration. The objective function value decreases in a monotone manner all the way but there are phase changes from batch to incremental and vice versa. Observe that the batch k-means algorithm converges to a local optimum after 8 iterations, generating the clustering with an objective function value 408,447 and the confusion matrix in the left panel of Table 1. Following the batch iterations are 20 incremental iterations, each of which moves a sample between clusters and decreases the objective function value a little bit. The 20 moves help the batch algorithm get out of the local

---

[2]Downloaded from http://www.chestsurg.org/microarray.htm
[3]Downloaded from http://wwwgenome.wi.mit.edu/mpr/data_set_ALL_AML.html

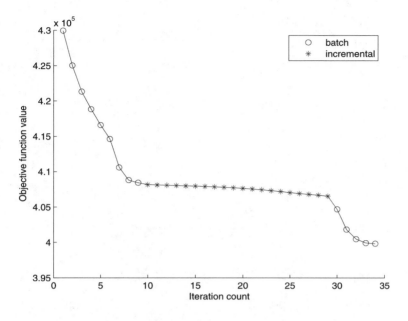

Fig. 1. Lung cancer data: Objective function value vs. iteration count

Table 1. Lung cancer: confusion matrix by batch k-means (left) and by Algorithm 1 (right)

Cluster #	ADCA	MPM
1	100	0
2	50	31

Cluster #	ADCA	MPM
1	150	0
2	0	31

optimum. So after the 20 incremental iterations, there are 5 more batch iterations, which substantially decrease the objective function value. Eventually we get the diagonal confusion matrix as in the right panel of Table 1 and the final objective function value is $399,821$.

In over 100 random runs on the leukemia data, on average, 25 samples are misclassified by the batch k-means but only 2 by Algorithm 1. Figure 2 shows a particular run of Algorithm 1 on leukemia data. As in the previous example, incremental iterations lead the batch k-means out of local minima and trigger more sample moves between clusters. Table 2 contains the confusion matrix when the batch k-means converges at the fifth iteration from the start and the confusion matrix when Algorithm 1 stops. The objective

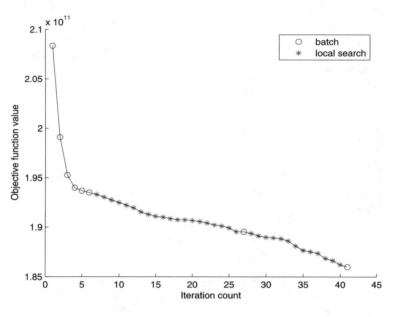

Fig. 2. Leukemia data: Objective function value vs. iteration count

Table 2. Leukemia: confusion matrix by batch k-means (left) and by Algorithm 1 (right)

Cluster #	ALL	AML
1	28	16
2	19	9

Cluster #	ALL	AML
1	47	2
2	0	23

function values at the fifth iteration is 1.94×10^{11} and the final objective function value is 1.86×10^{11}.

To exhibit the efficiency of our BIRCH-like approach, we compare it with the k-means algorithm using KL divergence, also known as the divisive information-theoretic clustering (DITC) algorithm [137], and its enhanced variants [130], on the *CMU newsgroup* text data. This data set consists of approximately 20000 newsgroup postings collected from 20 different usenet newsgroups (see Table 3). One thousand messages are in each of the 20 newsgroups. Some of the newsgroups have similar topics (e.g., comp.graphics, comp.os.ms-windows.misc, and comp.windows.x), while others are unrelated (e.g., alt.atheism, rec.sport.baseball, and sci.space). To preprocess the

Table 3. CMU newsgroup20: group names

Group #	Group name
1	alt.atheism
2	comp.graphics
3	comp.os.ms-windows.misc
4	comp.sys.ibm.pc.hardware
5	comp.sys.mac.hardware
6	comp.windows.x
7	misc.forsale
8	rec.autos
9	rec.motorcycles
10	rec.sport.baseball
11	rec.sport.hockey
12	sci.crypt
13	sci.med
14	sci.electronics
15	sci.space
16	soc.religion.christian
17	talk.politics.guns
18	talk.politics.mideast
19	talk.politics.misc
20	talk.religion.misc

documents and build the vector-space model [381], we first remove the header of each document, except for the subject line, then we use the Bow software [330] to filter out stopwords and select the 2,000 words with the highest contribution to the mutual information as features to represent documents. Documents with less than 10 word occurrences are discarded. After the pre-processing, we are left with a $2,000 \times 17,508$ word-document matrix. Then we apply the popular (term frequency inverse document frequency $tfidf$) scaling [380] on this matrix and normalize each document vector to have L_1 norm 1.

In DITC, each document is treated as a probability distribution over words and the objective function to minimize is

$$\sum_{l=1}^{k} \sum_{a \in \pi_l} p(a) KL(a, x^l),$$

where $a = (a_1, ..., a_n)$, $\sum_{j=1}^{n} a_j = 1$, $p(a)$ is the prior for a and $\sum_{a \in \mathcal{A}} p(a) = 1$, $KL(a, x^l)$ is the KL divergence between a and its cluster centroid $x^l = \sum_{a \in \pi_l} p(a)a / \sum_{a \in \pi_l} p(a)$. Generally, a word-document matrix is very sparse; in our case, more than 98% of the entries are zeros. Dhillon and Guan [130] find that DITC can falter in the presence of sparsity and propose the DITC_prior algorithm using an annealing factor to perturb the centroids to avoid infinite KL divergences and to improve the clustering results. At the beginning of DITC_prior the annealing factor is set to a large value, then it subsequently decreases every iteration and eventually becomes zero. Dhillon and Guan also study the performance of the merger of DITC and DICT_prior with their incremental version, which are referred to as DITC_merger and DITC_prior_merger in this section. DITC_prior is used in our BIRCH-like approach because it gives almost as good results as DITC_prior_merger but it is much faster than the latter [130].

We first apply PDDP on the word-document matrix and generate 100, 200, 500, and 1000 document clusters. Then for each case, we compute the cluster centers and then cluster the centers into 20 clusters using DITC_prior. After we get the clustering of the cluster centers, we project it to the clustering of the original documents, i.e., documents that are clustered into the same cluster by PDDP will have the same projected cluster membership. For comparison in clustering quality, we also use the projected clustering as initialization and run DITC and DITC_prior on the entire set of documents. We call this last strategy BIRCH-like+DITC and BIRCH-like+DITC_prior approaches.

Since we know the class label for each document, we could form the confusion matrix; however, instead of showing a 20×20 matrix, we introduce a scalar for cluster validity, *normalized mutual information* (NMI), which is computed based on the confusion matrix, as

$$\frac{2 \sum_{l=1}^{k} \sum_{h=1}^{c} \frac{n_{l}^{(h)}}{n} \log \left(\frac{n_{l}^{(h)} n}{\sum_{i=1}^{k} n_{i}^{(h)} \sum_{i=1}^{c} n_{l}^{(i)}} \right)}{H(\pi) + H(\zeta)},$$

where c is the number of classes, n_i is the number of documents in cluster i, and $n^{(j)}$ is the number of documents in class j, $H(\pi) = -\sum_{i=1}^{k} \frac{n_i}{n} \log \left(\frac{n_i}{n} \right)$, and $H(\zeta) = -\sum_{j=1}^{c} \frac{n^{(j)}}{n} \log \left(\frac{n^{(j)}}{n} \right)$. An NMI value is between 0 and 1, and a high NMI value indicates that the clustering and the true class label match well.

Tables 4–6 compare the NMI values, objective function values, and computation time by DITC (averaged over 10 random runs), DITC_merger (averaged over 10 random runs), DITC_prior, DITC_prior_merger, BIRCH-like, BIRCH-like+DITC, and BIRCH-like+DITC_prior approaches. Among all the algorithms, mergers are most computationally expensive. Between the BIRCH-like approaches and DITC with random initialization, the BIRCH-like approach takes almost the same time but gives better NMI values. As we increase the number of clusters generated by PDDP, the computation time increases because more centers need to be clustered; however, the highest NMI value is achieved for 1,000 clusters. The BIRCH-like+DITC_prior gives as good as or better clustering results, compared to DITC_prior and takes much less time. The performance of BIRCH-like+DITC is slightly worse than

Table 4. CMU newsgroup20: NMI values by DITC, DITC_merger, DITC_prior, and DITC_prior_merger applied to the entire newsgroups (top) and by BIRCH-like, BIRCH-like+DITC, and BIRCH-like +DITC_prior (bottom)

DITC	DITC_merger	DITC_prior	DITC_prior_merger
0.42	0.55	0.62	0.63

#clusters by PDDP	BIRCH-like	BIRCH-like+DITC	BIRCH-like+DITC_prior
100	0.45	0.57	0.63
200	0.46	0.58	0.62
500	0.47	0.59	0.62
1,000	0.5	0.6	0.64

Table 5. CMU newsgroup20: Objective function values by DITC, DITC_merger, DITC_prior, and DITC_prior_merger applied to the entire newsgroups (top) and by BIRCH-like, BIRCH-like+DITC, and BIRCH-like +DITC_prior (bottom)

DITC	DITC_merger	DITC_prior	DITC_prior_merger
4.82	4.73	4.56	4.55

#clusters by PDDP	BIRCH-like	BIRCH-like+DITC	BIRCH-like+DITC_prior
100	4.84	4.66	4.56
200	4.81	4.63	4.57
500	4.78	4.6	4.56
1,000	4.78	4.61	4.56

Table 6. CMU newsgroup20: computation time (in seconds) by DITC, DITC_merger, DITC_prior, and DITC_prior_merger applied to the entire newsgroups (top) and by BIRCH-like, BIRCH-like+DITC, and BIRCH-like +DITC_prior (bottom)

DITC	DITC_merger	DITC_prior	DITC_prior_merger
10	859	53	831

#clusters by PDDP	BIRCH-like	BIRCH-like+DITC	BIRCH-like+DITC_prior
100	9	15	31
200	10	16	31
500	12	18	33
1,000	14	20	37

DITC_prior but BIRCH-like+DITC runs much faster than DITC_prior. This is because the BIRCH-like approach is a good initialization, so the DITC_prior algorithm can use smaller starting annealing factor and thus converges in fewer iterations as shown in Table 7.

Table 7. CMU newsgroup20: iteration count of DITC, DITC_merger, DITC_prior, and DITC_prior_merger (top) and of BIRCH-like+DITC, and BIRCH-like+DITC_prior (bottom) applied to the entire newsgroups

DITC	DITC_merger	DITC_prior	DITC_prior_merger
9	29	25	45

#clusters by PDDP	BIRCH-like+DITC	BIRCH-like+DITC_prior
100	3	15
200	3	11
500	3	15
1,000	3	16

Acknowledgments

The research of Dhillon and Guan was supported in part by NSF grant CCF-0431257, NSF Career Award ACI-0093404, and NSF-ITR award IIS-0325116. The research of Kogan and Teboulle was supported in part by the United States–Israel Binational Science Foundation (BSF). Kogan's work was also supported by the Fulbright Program and Northrop Grumman Mission Systems.

Sampling Methods for Building Initial Partitions

Z. Volkovich, J. Kogan, and C. Nicholas

Summary. The initialization of iterative clustering algorithms is a difficult yet important problem in the practice of data mining. In this chapter, we discuss two new approaches for building such initial partitions. The first approach applies a procedure for selecting appropriate samples in the spirit of the Cross-Entropy (CE) method, and the second is based on a sequential summarizing schema. In the first approach, we use a sequential sample clustering procedure instead of the simulation step of the CE method. In this context, we state several facts related to the Projection Pursuit methodology for exploring the structure of a high-dimensional data set. In addition we review several external and internal approaches for cluster validity testing. Experimental results for cluster initializations obtained via the CE method and the first of the presented methods are reported for a real data set.

1 Introduction

The purpose of unsupervised classification, or cluster analysis, is to partition data (objects, instances) into similar groups so that objects in the same group are more similar than those from different groups. The similarity between the dataset elements is frequently described by means of a distance function. Computer-supported analysis based on this similarity function is intended to separate items into homogeneous sets, to provide an interpretation of the inner configuration of the observations, and to identify any "natural" group. Generally speaking, most existing clustering methods can be categorized into three groups: partitioning, hierarchical, and density-based approaches. The hierarchical procedures yield a nested sequence of partitions and, as a rule, avoid specifying how many clusters are appropriate. This sequence of partitions is often presented in the form of dendrograms, or tree diagram, and the desired number of clusters can be found by cutting the dendrogram at some level. One of the principal disadvantages of hierarchical algorithms is the high computational cost, making most of them unviable for clustering large datasets. Partitioning methods have the benefit of being able to incorporate knowledge about the size of the clusters by using certain templates and the elements' dissimilarity in the objective function. Such a procedure is guaranteed to produce clustering for any data; however, it is very hard to achieve the global optimal partition.

In general, clustering algorithms make different assumptions about the dataset structure. Well-known iterative procedures, like k-means clustering or k-median clustering, assume that the analyzed set consists of a number of separated subsets of data points such that each one is suitably (spherically in the case of k-means) distributed around its center. Since for the cases of interest one does not know whether these assumptions are satisfied by the data, a final partition of the k-means iteration process essentially depends on an initial partition of a dataset. How "good" initial partitions can be constructed remains an open question. Employing sampling procedures for statistical estimation of initial centroids can be considered as a special case of the sampling-base meta-algorithms suggested in [92] and [392]:

1. Draw a random sample from the underlying dataset,
2. Find a good clustering of the sample,
3. Extend the clustering of the sample to a clustering of the full dataset.

The significant point in converting these models into a specific algorithm is the realization of the (third) extension step. There are at least two options in the case of the k-means like clustering algorithms. The first approach is to use collected sample centroids as initial centroids for entire dataset clustering. The second (naïve) approach defines the final partition by assigning each point to the nearest sample centroid.

We note that different sampling techniques have been proposed in the past, for example, speeding up performance of the classical k-means [122], or finding the "true" number of clusters [148].

In this chapter, in Sects. 5 and 6 we discuss two new approaches for building an initial cluster partition based on a bootstrapping technique in conjunction with the Gaussian Mixture Model of data, comparing it to the well-known Cross-Entropy (CE) method [376]. One of the approaches applies a procedure for selecting appropriate samples in the spirit of the CE method and the other is based on a sequential summarizing schema. In the first approach we use a sequential sample clustering procedure instead of the simulation step of the CE method.

First, we state in Sect. 2 the well-known clustering Expectation-Maximization (EM) and k-means models, and related statistical facts. A difficulty in applying sample procedures for multivariate data is the problem of "the curse of dimensionality" (a term coined by Bellman [54]). The curse of dimensionality means that most high-dimensional projections are empty unless the sample is quite large. This is often the case with the sparse data that are typical in text mining applications (see, for example, [60]). In Sect. 3, we review several statistical techniques for projection index pursuit that make it possible to overcome this problem by initializing clusters in appropriate "interesting" low-dimensional subspaces.

Section 4 is devoted to several internal and external approaches for cluster validation testing. Experimental results comparing cluster initializations obtained via the CE method and the first of the presented methods are reported for real data in Sect. 7.

2 Clustering Model

Let $\{x_1, ..., x_m\}$ be a set of vectors in a subset X of the n-dimensional Euclidean space R^n. Consider a partition $\Pi = \{\pi_1, \ldots, \pi_k\}$ of the set, i.e.,

$$\pi_1 \cup \cdots \cup \pi_k = \{x_1, \ldots, x_m\} \quad \text{and} \quad \pi_i \cap \pi_j = \emptyset \text{ if } i \neq j.$$

Given a real valued function q whose domain is the set of subsets of $\{x_1, \ldots, x_m\}$ the quality of the partition is given by $Q(\Pi) = q(\pi_1) + \cdots + q(\pi_k)$. The problem is to identify an optimal partition $\{\pi_1^o, \ldots, \pi_k^o\}$, i.e., a partition that optimizes $Q(\Pi)$. Often the function q is associated with a "dissimilarity measure," or a "distance-like" function $d(x, y)$. We refer to $d(\cdot, \cdot)$ as a "distance-like" function, since d is not required to be symmetric or satisfy the triangle inequality. Use of such a function commonly provides better grouping in comparison with classical dissimilarity measures like the squared standard Euclidean distance [278, 279, 432]. However, we restrict ourselves to considering only the squared standard Euclidean distance function, since the choice of distance function does not make much difference in the case of relatively small samples. The above-mentioned optimization problem is represented in this case as:

$$\min_{c_j} R(c_1, ..., c_k) = \sum_{j=1}^{k} \sum_{z \in X} \min_{c_j} \|z - c_j\|^2. \tag{1}$$

The k-means method produces an approximate solution to this optimization problem in an iterative way:

1. Construct an initial partition.
2. Minimization: calculate the mean (centroid) of each cluster's points.
3. Classification: assign each element to the nearest current centroid.
4. Repeat Steps 2 and 3 until the partition is stable, that is, until the centroids no longer change.

The convergence proof of the k-means clustering algorithms requires showing two easy facts:

- Reassigning a point to a different group does not increase the error function;
- Updating the mean of a group does not increase the error function.

However, k-means often leads to a partition consisting of the so-called "nonoptimal stable clusters." One way to overcome this problem is to use the incremental k-means algorithm (see, e.g., [132, 133, 144, 280]). Note that this incremental k-means algorithm gives more accurate results in the case of relatively small clusters and is frequently able to escape local minima. The k-means approach can be considered as a simplification of the well-known EM algorithm.

The EM algorithm [100, 173, 464] is an iterative procedure for finding the maximum-likelihood estimate of the parameters for the mean and variance–covariance of each group, and the mixing proportion in situations when the number of components k is known. Recall that this algorithm fits the Gaussian Mixture Model (GMM) of data, that is, the underlying probability density of X is represented by

$$f(x) = \sum_{i=1}^{k} p_i G(x|\mu_i, \Sigma_i),$$

where $G(x|\mu, \Sigma)$ is the Gaussian density with mean μ and covariance matrix Σ. In the special spherical case

$$\Sigma_i = \sigma^2 I, \ i = 1, \ldots, k,$$

where I is the $n \times n$ identity matrix and σ^2 is the unknown dispersion of each cluster, the standard k-means algorithm can be shown to be a version of the EM algorithm and each component $\Pi = \{\pi_1, \ldots, \pi_k\}$ is parametrized by its average μ_i, $i = 1, \ldots, k$ associated with the cluster centroid $c_j, j = 1, \ldots, k$. That is, the clusters are spherical, centered at the means μ_i, $i = 1, \ldots, k$.

Given a partition $\Pi = \{\pi_1, \ldots, \pi_k\}$, $2 \le k$, the optimization criterion presented in (1) is based on the total dispersion matrix or total scatter matrix:

$$T_k = \sum_{j=1}^{k} \sum_{z \in \pi_j} (z - \overline{\mu})(z - \overline{\mu})^{\mathrm{t}}, \tag{2}$$

where $\overline{\mu}$ is the arithmetic mean of the set X. Matrices B_k and W_k of between and within k-clusters sums of squares are defined as

$$B_k = \sum_{j=1}^{k} |\pi_j| (\overline{\mu}_j - \overline{\mu})(\overline{\mu}_j - \overline{\mu})^{\mathrm{t}}, \quad W_k = \sum_{j=1}^{k} \sum_{z \in \pi_j} (z - \overline{\mu}_j)(z - \overline{\mu}_j)^{\mathrm{t}}, \tag{3}$$

where $\overline{\mu}_j$ is the arithmetic mean of π_j, $j = 1, ..., k$, and $T_k = W_k + B_k$ (see, for example, [326]).

Actually, the optimization problem corresponds to minimizing the trace of W_k. Note that in the framework of GMM, the distributions of T_k, W_k, and B_k can be calculated analytically under the null hypothesis that there are "no clusters" in the data, i.e., the data are one big cluster. Estimates of cluster validity based on these statistics have been developed [96, 222, 296].

3 Projection Pursuit

Projection pursuit is a statistical methodology for exploring the structure in high-dimensional datasets by means of low-dimension projectors. Projection

pursuit works by optimizing a predetermined criterion function, called a projection pursuit index. Principal component analysis (PCA) is one of the more
traditional tools used to create an informative low-dimensional data representation. However, PCA does not necessarily provide the "best" representation
of the data and may miss "interesting" details of the data structure. Hence,
much research has been done on methods that identify projections having the
ability to display most "interesting" features of the data.

It appears that the idea of a projection pursuit index first appeared in the
papers of Kruskal [295] and Friedman and Tukey [178]. The projection onto
an appropriate "interesting" low-dimensional subspace often makes it possible to overcome the aforementioned "curse of dimensionality." An interesting
low-dimensional subspace is chosen according to a projection pursuit index
optimization. From this point of view, PCA can be considered a special case
of the projection pursuit procedure for which the index is the variance of data
maximized over all unit projections.

Many indices have been developed from different standpoints. Supervised classification applications (see, for example, [80, 306]) commonly employ indices based on maximization of the relationship between-group sums
of squares and within-group sums of squares (see the definition in Sect. 2).
Unsupervised classifications take advantage of indices based on a so-called
"departure from normality." As shown by Diaconis and Freedman [138], under
some weak assumptions, distributions of linear projections can be considered
as approximately normal in the high-dimension case, i.e., practically speaking,
most projections are approximately Gaussian distributed. By the well-known
Cramer–Wold principle, if all one-dimensional projections are normal then the
underlying distribution is normal too. However, a particular linear projection
may not let us see the clustering structure in a high-dimension dataset, even
if some other projection would show us the peaks in projection density that
correspond to the clusters in the dataset.

Another simple interesting interpretation of this concept has been discussed in [79]. For a given dataset the index corresponding to PCA is

$$V = \max_{a}\langle a, Sa\rangle, \tag{4}$$

where $a \in R^n$ is a direction, S is the sample covariance matrix, and $\langle a, b\rangle$ is
the scalar product of $a, b \in R^n$. It is well known that a vector a maximizing
$\langle a, Sa\rangle$ is the eigenvector corresponding to the largest eigenvalue of S, i.e.,
a is the direction of maximum variance in the data. We suppose that the
underlying distribution is normal with the parameters μ and $\sum$. The log-
maximum likelihood function of the direction a is

$$l(a) = -\frac{mn}{2}\left(1 + \log\left(\frac{2\pi m}{m-1}\right)\right) - \frac{m}{2}\log(V), \tag{5}$$

where m is the dataset size and n is the data dimension. For all a this function is maximized with the same estimators of the parameters, hence this

function can be viewed as a pursuit index. Moreover, this function decreases monotonically as a function of V. Then, if normality is assumed, the most interesting projections correspond to the smallest values of $l(a)$, i.e., the most interesting projection possesses the smallest likelihood values.

Now we describe several widespread indexes. A review of dimension reduction techniques including such indexes can be found in [98].

3.1 Friedman's Pursuit Index

Friedman's Pursuit Index [175] is intended to search for those directions a in which the marginal distribution based on the projection has the maximum departure from the univariate Gaussian law. For a given a, the projection $Y = \langle a, X \rangle$ is mapped into $[-1, 1]$ by the formula

$$p_r(x) = \frac{\frac{1}{2} p_Y (\Phi^{-1}(\frac{x+1}{2}))}{\varphi(\Phi^{-1}(\frac{x+1}{2}))}, \tag{6}$$

where p_Y is the density of Y, φ is the standard normal density, and Φ is the cumulative normal standard distribution. Hence, the purpose is to find a direction such that the density $p_r(x)$ is most different from $\frac{1}{2}$ in the L_2 norm, i.e., to maximize the functional

$$l(a) = \int\limits_{-1}^{1} \left(p_r(x) - \frac{1}{2} \right)^2 \mathrm{d}x = \int\limits_{-1}^{1} p_r^2(x) \, \mathrm{d}x - \frac{1}{2}. \tag{7}$$

For the calculation of this index we approximate the function $p_r(x)$ by a series of the Legendre polynomial:

$$p_r(x) = \sum_{k=0}^{\infty} c_k L_k(x), \tag{8}$$

where

$$c_k = \frac{2k+1}{2} \int\limits_{-1}^{1} L_k(x) p_r(x) \, \mathrm{d}x. \tag{9}$$

Substituting this into the index expression above, we have

$$l(a) = \sum_{k=0}^{\infty} \frac{2k+1}{2} E(L_k(x))^2 - \frac{1}{2},$$

where the expectation E is with respect to the random variable x. This value can be estimated using empirical expectation as:

$$\hat{E}(L_k(x)) = \frac{1}{m} \sum_{i=0}^{m} L_k(2\Phi(\langle a, X_i \rangle) - 1). \tag{10}$$

Recall that m is the size of the dataset being considered. The required calculation can be performed very quickly using the recurrence relationship:

$$L_0(x) = 1, \; L_1(x) = x,$$

$$L_k(x) = \frac{2k-1}{k} x L_{k-1}(x) - \frac{k-1}{k} L_{k-2}(x).$$

The series expansion in the expression for $l(a)$ has to be truncated to avoid overfilling, since the high-order estimated coefficients are unstable even for large samples. Typically, no more than eight terms in the expansion are considered. On the other hand, if the sample size is large enough, then we can start from small values of the terms and later improve local maxima by increasing the number of terms.

As shown in [112] the Friedman index can be rewritten as

$$l(a) = \int\limits_{-\infty}^{\infty} \frac{(f(x) - \varphi(x))^2}{2\phi(x)} \, dx, \tag{11}$$

where $f(x)$ is the normalized projection density. Hence, this index is a special case of a general set of indexes based on the orthogonal polynomial considered in [112]. For instance, the Hall index [214] is

$$l_H(a) = \int\limits_{-\infty}^{\infty} (f(x) - \varphi(x))^2 \, dx, \tag{12}$$

and the natural Hermit index is

$$l_C(a) = \int\limits_{-\infty}^{\infty} (f(x) - \varphi(x))^2 \, \phi(x) \, dx. \tag{13}$$

All these can be computed via an orthogonal expansion of $f(x)$ using Hermite polynomials, where $\phi(x)$ is (as usual) the density of the standard normal distribution.

3.2 Entropy-Based Indexes

Another way to measure departure from normality is by using entropy (sometimes also called differential entropy) [235, 252]. It is well known that the entropy H of a random vector having density f is:

$$H = -\int f(x) \log(f(x)) \, dx. \tag{14}$$

The natural information-theoretic one-unit contrast function is negentropy, defined as

$$J(f) = H(p_g) - H(f), \qquad (15)$$

where p_g is a Gaussian density with mean and covariance equal to those of f. Several properties of entropy related to the pursuit index concept have been discussed in [252]. In particular, Gibb's second theorem states that a multivariate Gaussian density maximizes the entropy with respect to f over all distributions with the same mean and covariance. For any other non-normal distribution, the corresponding entropy is strictly smaller. Therefore, negentropy will always be positive for non-normal distributions and will be "0" if the distribution f is Gaussian. The estimation of entropy is rather problematic in practice. Two ways to evaluate the entropy may be suggested:

- Replace the integral with its sample value by numerical integration;
- Approximate by means sample statistics and univariate nonparametric density estimations.

However, these approaches are recognized as computationally ineffective and theoretically difficult. Simpler estimates of entropy are frequently based on the polynomial expansions of the projection density or on the Gram–Charlier–Edgeworth approximations. For example, negentropy can be approximated for standardized random values by higher-order cumulants as follows:

$$J \simeq -\frac{1}{12}\left(k_3^2 + \frac{1}{4}k_4^2\right) \qquad (16)$$

where $k_i, i = 3, 4$, are the ith order cumulants. As argued in [237], cumulant-based approximations appear to be inaccurate and too sensitive to outliers. Approximations of negentropy suggested in [237] can be represented in the simplest case as:

$$J \simeq c\left(E(V(x)) - E(V(g))\right), \qquad (17)$$

where V is any nonquadratic function, c is an arbitrary constant, and g is a standardized Gaussian variable. Recall that x is assumed to be standardized as well. For $V(x) = x^4$, we get the modulus of kurtosis. The statistical characteristics of these estimators were analyzed as well. It was shown that for an appropriate selection of V the asymptotic variance and robustness are significantly better than analogous cumulant-based estimators. The following options for V were considered:

$$V_1 = \log(\cosh(a_1 x)), \quad V_2 = \exp\left(-\frac{a_2 x^2}{2}\right). \qquad (18)$$

Experimentally, it was found that especially the values $1 \le a_1 \le 2$ and $a_2 = 1$ provide good approximations.

3.3 BCM Functions

As mentioned in Sect. 3.2, computationally attractive projection indices based on polynomial moments are not directly applicable, as they heavily emphasize departure from normality in the tails of the distribution. Friedman tried to overcome this problem by using a nonlinear projection on the data on $[-1, 1]$ (as described earlier). On the other hand, the Friedman index is insensitive to multimodality in the projected distribution in the case of significant difference distinction between the picks' sizes. The insensitivity is caused by the L^2 norm approximation.

An approach for exploring the projection multimodality has been considered in the framework of the synaptic modification theory of Bienenstock, Cooper, and Munro neurons (BCM). BCM yields synaptic modification equations that maximize a projection index $l(a)$ as the function of a direction a. In this context, $l(a)$ is called a cost function and it measures the deviation from a Gaussian distribution. Synaptic modification equations are achieved using gradient ascent with respect to the weights (see, for example, [71]).

Intrator and Cooper [239] introduced a cost function to assess the deviation from a Gaussian distribution in the form of multimodality:

$$l(a) = \frac{1}{3}E(\langle a, X \rangle^3) - \frac{1}{4}E^2(\langle a, X \rangle^2)$$

As was noted in [71], the cost function

$$l(a) = E\left[\langle a, X \rangle^2 \left(1 - \frac{1}{2}\langle a, X \rangle^2\right)\right],$$

can be regarded as a cost function for PCA. Other cost functions based on skewness and kurtosis of a projection were also discussed.

4 Cluster Validation

In general, validating clustering solutions means evaluating results of cluster analysis in a quantitative and objective manner. Such an evaluation can be built on two kinds of criteria:

- Internal criteria: the quality measure is fully based on the data themselves,
- External criteria: a clustering solution is based on a prior information, i.e., external information that is not contained in the dataset. For instance, it can be a classification obtained by another approach or preprocessed information about data structure.

In this section we recall several facts regarding the internal and external criteria. In this presentation we partly follow [145].

4.1 Internal Criteria

Several methods have been proposed for inner testing of cluster validation. Jain and Dubes [242] provided an overview of statistical validation approaches based on testing the null hypothesis in the absence of clustering together with estimating the number of cluster component in a set. As usual, these procedures are based on the within-clusters, and possibly between-clusters, sums of squares described in Sect. 2. Such methods employ the same observations that are used to create the clustering. Consequently, the distribution of suitable statistics is intractable. In particular, as clustering routines attempt to maximize the separation between clusters, ordinary significance tests such as analysis of variance F-tests are not valid for testing differences between the clusters. A Monte Carlo evaluation of 30 internal indices was provided in [457].

We state here several inner procedures. Recall that for a dataset partitioned into k clusters, define B_k and W_k to be the matrices of between and within clusters sums of squares (see Sect. 2). The following indices are frequently used to estimate the number of clusters in a dataset.

1. The Calinski and Harabasz index [96] defined as

$$CH_k = \frac{\operatorname{tr}(B_k)/(k-1)}{\operatorname{tr}(W_k)/(m-k)}.$$

   The estimated number of clusters is given by k providing the maximum of CH_k, where m is the size of the considered dataset.

2. The Krzanowski and Lai index [296] defined by the following relationships

$$\operatorname{dif} f_k = (k-1)^{2/n} \operatorname{tr}(W_{k-1}) - k^{2/n} \operatorname{tr}(W_k)$$

$$KL_k = |\operatorname{dif} f_k|/|\operatorname{dif} f_{k+1}|.$$

   The estimated number of clusters matches the maximal value of the index KL_k, where n is the dimension of the dataset being considered.

3. The Hartigan index [462] defined as

$$h_k = \left(\frac{\operatorname{tr}(W_k)}{\operatorname{tr}(W_{k+1})} - 1 \right) (m - k - 1)$$

   The estimated number of clusters is the smallest $k \geq 1$ such that $h_k \leq 10$, where again m is the size of the dataset being considered.

4. Sugar and James [414] proposed an information-theoretic approach for finding the number of clusters in a dataset that can be considered an index in the following way. We select a transformation power t. A typical value is $t = n/2$, where n is the dimension of the considered dataset. Then calculate the "jumps" in transformed distortion,

$$J_k = \left(\operatorname{tr}(W_k)^{-t} - \operatorname{tr}(W_{k-1})^{-t} \right).$$

   The estimated number of clusters matches the maximal value of the index J_k.

5. In the Gap index [420], for each $k \geq 1$, compute the values $\text{tr}\,(W_k)$. B ($B = 10$ in the paper) reference datasets are generated under the null distribution assumption. Each of the datasets is presented to a clustering procedure, and the values $\text{tr}\left(W_k^1\right), ..., \text{tr}\left(W_k^B\right)$ are evaluated. The estimated gap statistics are calculated as

$$\text{gap}_k = \frac{1}{B} \sum_b \log\left(\text{tr}\left(W_k^b\right)\right) - \log\left(\text{tr}\left(W_k\right)\right).$$

Let sd_k be the standard deviation of $\log\left(\text{tr}\left(W_k^b\right)\right)$, $1 \leq b \leq B$, and

$$\hat{\text{sd}}_k = \text{sd}_k \sqrt{1 + \frac{1}{B}}.$$

The estimated number of clusters is the smallest $k \geq 1$ such that

$$\text{gap}_k \geq \text{gap}_{k^*} - \hat{\text{sd}}_{k^*},$$

where $k^* = \text{argmax}_{k \geq 1}(\text{gap}_k)$. The uniform distribution is chosen as a reference null distribution and two approaches are considered for constructing the region of support for the distribution (for details see [420]).

Many papers are devoted to identifying the possible number of clusters by means of resampling procedures. We point out only a few recent contributions, namely [311], and the Clest method [145, 179]. Indeed, such an approach is based on the "cluster stability" concept as it was summarized and generalized in [374, 375]. We state here an algorithm for the internal index computation presented in these works.

1. Split the dataset of size $2t$ into two sets of equal size, O_1^t and O_2^t.
2. Present the first dataset to the algorithm. The result is the mapping α_1 of each of the objects in O_1^t to one of k clusters.
3. Present the second dataset O_2^t to the algorithm. The result is the mapping α_2 of each of the objects in O_2^t to one of k clusters. Use α_2 to predict the cluster membership of all objects contained in the first set.
4. Set O_1^t now has two different labels. Find the correct permutation of labels by using the well-known Hungarian method for minimum weighted perfect bipartite matching (see [297]). The costs for identifying labels i and j are the number of misclassifications with respect to the labels α_1, which are assumed correct.
5. Normalize with respect to the random stability.
6. Iterate the whole procedure from Step 1 to Step 5, average over assignment costs, and compute the expected (in-)stability value.
7. Iterate the whole procedure for each k to be tested.

An application of the Hungarian method is required since the labels of the clusters can be permuted. A match between the labels can be found by solving

$$\phi(\alpha_1, \alpha_2) = \min_\pi \frac{1}{t} \sum_{j=1}^{t} 1\{\alpha_1(o_j) - \pi(\alpha_2(o_j))\},$$

where $1\{\cdot\}$ is an indicator function and π is a permutation of k possible labels. The Hungarian method has computational complexity of $O(k^3)$.

Normalization with respect to the random stability means:

$$\psi_k = \frac{\text{mean}(\phi_k(\alpha_1, \alpha_2))}{\text{mean}(\phi_k(\rho_1, \rho_2))}, \tag{19}$$

where ρ_1, ρ_2 are the random predictors, which assign labels uniformly at random. The estimated number of clusters is given by k providing the maximum of ψ_k.

It has been noted by the authors that splitting the total set of objects into two disjoint subsets is recommended because the size t of the individual sets should be large. However, this approach can be applied formally for any sample size.

4.2 External Criteria

We state in this part several external criteria. A formal way to do it is to exploit external indices of partition agreement. Usually, the calculation of these scores is based on the so-called crosstabulation, or contingency tables. Entries m_{ij} of this table denote the number of objects that are in both clusters i and j, $i = 1, ..., r$, $j = 1, ..., c$ for two different partitions P_r and P_c. Note that a relationship between two partitions can be considered in the framework of the measures of association for nominal data. The most well known of these is the Cramer correlation coefficient defined in the following way. Let us introduce

$$m_i^{(r)} = \sum_{j=1}^{c} m_{ij}, \ i = 1, ..., r, \quad m_j^{(c)} = \sum_{i=1}^{r} m_{ij}, \ j = 1, ..., c, \tag{20}$$

$$m = \sum_{i=1}^{r} m_i^{(r)} = \sum_{j=1}^{c} m_j^{(c)}; \tag{21}$$

the row and column sums of the contingency table, and the chi-square statistics

$$\chi^2 = \sum_{i=1}^{r} \sum_{j=1}^{c} \frac{(m_{ij} - e_{ij})^2}{e_{ij}}, \ e_{ij} = \frac{m_i \cdot m_j}{m}. \tag{22}$$

The Cramer coefficient is

$$V = \sqrt{\frac{\chi^2}{m \cdot \min(r - 1, c - 1)}}. \tag{23}$$

Several known external indexes employ the statistics

$$Z = \sum_{j=1}^{c} \sum_{i=1}^{r} m_{ij}^2. \tag{24}$$

Rand [366] introduced the index

$$R = 1 + \left(\frac{Z - 0.5 \cdot \left(\sum_{j=1}^{c} \left(m_j^{(c)} \right)^2 + \sum_{i=1}^{r} \left(m_i^{(r)} \right)^2 \right)}{\binom{man}{2}} \right).$$

Jain and Dubes [242] considered the index

$$JD = \frac{(Z - m)}{\left(\sum_{j=1}^{c} \left(m_j^{(c)} \right)^2 + \sum_{i=1}^{r} \left(m_i^{(r)} \right)^2 - Z - m \right)},$$

and Fowlkes and Mallows [169] provided the following expression

$$FM = \frac{(Z - m)}{2 \cdot \sqrt{\sum_{j=1}^{c} \binom{m_j^{(c)}}{2} \sum_{i=1}^{r} \binom{m_i^{(r)}}{2}}}.$$

It is easy to see that the two indexes R and FM are linear functions of Z, and for that reason each one is a linear function of the other. Often indexes are standardized such that an appropriate expected value is 0 when the partitions are selected at random and 1 when they match perfectly.

5 Sampling Approaches for Selection of Initial Centroids for k-Means Clustering Algorithms

In this section we discuss two methods for building an initial cluster partition based on bootstrapping and simulation techniques. The first method is an application of the CE method. The second method applies a procedure for selecting appropriate samples in the spirit of the CE method. This new approach uses a sequential samples clustering procedure instead of the simulation step of the CE method.

5.1 The Cross-Entropy Method in Clustering Problems

The CE method has been introduced as a new genetic approach to combinatorial and multiextremal optimization and rare event simulation and has found many applications in different areas (see the CE site[1]). Generally, this method consists of the following two steps:

---

[1] http://iew3.technion.ac.il/CE/about.php

1. Generate a sample of random data that fit parameters of the underlying distribution;
2. Update the parameters in order to produce a "better" sample in the next iteration.

In this section we recall how the CE technique can be applied to clustering under the assumption of the GMM (as suggested in [294]) under conditions described in Sect. 2. First we pick two numbers:

- N, the number of simulated samples, and
- ρ, the proportion of correct ("elite") samples.

In our case, the parameter σ^2 of the GMM is assessed as

$$\sigma^2 = \frac{1}{n} \sum_{i=1}^{n} \sigma^2(x_i),$$

where $\sigma^2(x_i)$ is the dispersion of the ith coordinate throughout the entire set X.

The procedure is as follows:

1. Choose, deterministically or randomly, the initial means $\mu_j, j = 1, \ldots, k$ and set the counter $t = 1$.
2. Generate N sequences of k potential independent centroids $\{Y_i = y_{ji}, j = 1, \ldots, k\}$, $i = 1, \ldots, N$ according to the normal laws $y_{ji} \sim N(\mu_j, \sigma^2 I)$ and calculate the corresponding objective function values R_i, $i = 1, \ldots, N$.
3. Rank the sequence R_i, $i = 1, \ldots, N$ and select $N^{\text{elite}} = [\rho N]$ "elite" samples corresponding to N^{elite} smallest values of R_i, with $\{Y_i, i = 1, \ldots, N^{\text{elite}}\}$ centroids (we reindex the centroids as needed). The parameters are updated as

$$\mu_j^{(t)} = \frac{1}{N^{\text{elite}}} \sum_{i=1}^{N^{\text{elite}}} y_{ji}, \quad \sigma^2 = \frac{1}{n} \sum_{j=1}^{n} \sum_{i=1}^{N^{\text{elite}}} \sigma^2(y_{ji}). \tag{25}$$

4. If the stopping criterion is met, then stop, else set $t = t + 1$ and go to Step 2.

As proposed in the CE literature, smoothed weighted updating or, the so-called component-wise updating can be applied as an alternative method in Step 3. There are various stopping criteria. For instance, the CE algorithm can be terminated when the values of the objective function stabilize to within a suitable tolerance. The method has been shown to be robust with respect to the choice of initial centroids [294]. The CE method has several disadvantages related mainly to the second simulation step:

1. In the case of a high-dimensional dataset a simulation procedure is computationally expensive.

2. All simulations are performed assuming that the underlying distribution can be properly approximated by means of the Gaussian distribution. The assumption is rarely satisfied for real data, which are often sparse, as in the case of Text Mining applications (see for instance [60]). Evidently, modeling of such a high-dimensional dataset by means of a mixed normal law can lead to a big deviation from the underlying distribution. This is an aspect of the aforementioned "curse of dimensionality."

The method we now introduce does not simulate data by normal distribution.

5.2 Sampling Clustering Algorithm

Now we describe the Sampling Clustering Algorithm. As in the CE approach we start with outlining the parameters:

- N, the number of drawn samples;
- M, the sample size;
- ρ, the proportion of correct ("elite") samples.

The algorithm consists of the following steps:

1. Take N random samples $S_j, j = 1, ..., N$ of size M from the dataset, and set counter $t = 1$.
2. Apply the k-means algorithm to partition the drawn samples and get new centroids $Y_i = \{y_{ji}, j = 1, ..., k\}$, $i = 1, .., N$ (initial k-means centroids are chosen randomly for $t = 1$ and are taken from the previous iteration for $t > 1$).
3. Calculate the objective function values R_i, $i = 1, ..., N$ for the partition obtained in Step 2.
4. Rank the sequence R_i, $i = 1, ..., N$ and take $N^{\text{elite}} = [\rho N]$ "elite" samples corresponding to the smallest $[\rho N]$ values of R_i, say, $S_1, ..., S_{N^{\text{elite}}}$.
5. Update the centroids using the k-means algorithm for clustering the combined sample

$$\hat{S}_t = \bigcup_{i=1}^{N^{\text{elite}}} S_i.$$

   Initial k-means centroids are chosen randomly for $t = 1$ and are taken from the previous iteration for $t > 1$.
6. If the stopping criterion is met, then stop and accept the obtained centroids as an estimate for the true centroids, otherwise set $t = t + 1$ and go to Step 2.

We denote k centroids corresponding to a partition $\hat{S}_t$ by $\{c_1^t, ..., c_k^t\}$. The stopping criterion is based on convergence of the set $\{c_1^t, ..., c_k^t\}$ as $t \to \infty$. However, it should be clear that c_1^{t+1} and c_1^t do not necessarily correspond to the same centroid in the optimal partition of the dataset. This problem can be overcome by means of a minimization procedure for actual misclassifications between two sequential steps over all possible label permutations Ψ. Specifically, let $\hat{S}_t$ and $\hat{S}_{t+1}$ be two samples. We label each element of $\hat{S}_{t+1}$ by the nearest centroid of $\hat{S}_t$ and denote it as $\alpha_t(\hat{S}_t)$. A preferred label permutation ψ^* is chosen as a permutation, which leads to the smallest misclassification between $\alpha_t(\hat{S}_t)$ and the permuted source tagging, i.e.,

$$\psi^* = \arg\min(\psi \in \Psi \mid D(\alpha_t(\hat{S}_{t+1}), \psi(\alpha_{t+1}(\hat{S}_{t+1})),$$

where D assigns the number of misclassifications to two partitions. Note that we do not need to test all $k!$ possible permutations, because this problem can be represented as a special case of the minimum weighed perfect bivariate matching problem. Due to the well-known Hungarian method (see [297]) the computational complexity of this problem is $O(k^3)$. A stopping criterion can be specified by the distortion between centroids in step t and permuted by means of ψ^* at step $t+1$.

6 Iterative Averaging Initialization

In this section we briefly state an iterative clustering initialization method similar to that described in our earlier paper [432]. We outline the following parameters:

- N, the number of drawn samples;
- M, the sample size.

The centroids generating procedure is given next.

1. Choose randomly k elements $c_j^{(0)}$, $j = 1, ..., k$ of the dataset and set $t = 1$ (level counter).
2. Draw sample S_t of size M from the dataset, and set $t = t + 1$.
3. Apply the k-means algorithm with initial centroids $c_j^{(t-1)}$, $j = 1, ..., k$ to partition the sample S_t and obtain new centroids $c_j^{(t)}$, $j = 1, ..., k$.
4. If $t > M$ then stop, else goto step 2.

The initial centroids for the k-means clustering algorithm applied to the entire dataset are the average of the corresponding final centroids generated for all M samples.

6.1 Representative Samples

We consider random samples S_j, $j = 1, \ldots, N$ with replacement. For a given sample S we denote by s_i the cardinality of $S \cap \pi_i$, i.e., $s_i = |S \cap \pi_i|$. We denote $(s_1 + \ldots + s_k)$ by s and $\min\{s_1, \cdots, s_k\}$ by s_0. Due to the assumption on disjoint clusters, the mutual distribution of the variables $s_1, \ldots, s_k$ is a multinomial, i.e.,

$$P(s_1 = v_1, \ldots, s_k = v_k) = \frac{s!}{v_1! \cdots v_k!} p_1^{v_1} \cdots p_k^{v_k}, \tag{26}$$

which implies the binomial distribution $Bin(s, p)$ with parameter $p = p_i$ for each variable s_i

$$P(s_i = v) = \frac{s!}{v!(s-v)!} p_i^v (1 - p_i)^{n-v}, \; v = 0, 1, \ldots, s. \tag{27}$$

In what follows we focus on the binomial model. We use the normal approximation to the binomial distribution (see, for example, [160], part 7). The binomial distribution $Bin(s, p)$ with parameters s and p can be approximated by the normal distribution with mean $E(X) = sp$ and the standard deviation $\sigma = \sqrt{sp(1-p)}$ for large s (e.g., $s > 20$), and p bounded away from 0 and 1 (for example, $0.05 < p < 0.95$). One of the approximations can be described as follows (see [292], p. 101):

$$9 < sp(1 - p), \quad \frac{1}{s+1} < p < \frac{s}{s+1}, \quad \text{and } 1.07 < sp^{3/2}, \tag{28}$$

then for each x we have

$$\left| F(x) - \Phi\left(\frac{x + 0.5 - sp}{\sqrt{sp(1-p)}} \right) \right| \leq 0.05 \tag{29}$$

where Φ is the cumulative distribution function of the standard normal distribution and $F(x)$ is the commutative distribution function of $Bin(s, p)$. This approximation can be obtained from the convergence rate of the central limit theorem for the binomial distribution. For $p = p_0$ we choose the sample size s to satisfy the inequalities in (28). The choice of s yields (29) for all x and all p_i, $i = 1, \ldots, k$.

While sampling we will consider centroids of sample clusters as random variables. Since a centroid is an arithmetic mean of a cluster's elements, the variance of the centroids' coordinates is bounded from above by σ^2/s_0. To ensure a high probability P (say $P \geq 0.9$) to obtain samples as described above we set $F(s_0) \leq 1 - P$. Due to (29) this inequality yields

$$\Phi\left(\frac{s_0 + 0.5 - sp_0}{\sqrt{sp_0(1-p_0)}} \right) \leq 0.95 - P, \text{ and } s_0 + 0.5 - sp_0 = \gamma_P \cdot \sqrt{sp_0(1-p_0)}. \tag{30}$$

Here γ_P is a quintile of the standard normal distribution $\Phi(\gamma_P) = 0.95 - P$ (i.e., $\gamma_P = -1.65$, if $P = 0.9$). The lower bounds $\underline{s}$ for the sample size s with $s_0 = 10$, $P = 0.9$, $\gamma_P = -1.65$, for a variety of values for p_0 follow from (28) and (30) and are given in Table 1.

6.2 Resampling

For a given set of centroids M we generate the initial centroids for the k-means like algorithm as the arithmetic mean of the sample centroids. We

Table 1. Bounds $\underline{s}$ for the sample size

p_0	$9/p_0(1 - p_0)$	$1.07/p_0^{\frac{3}{2}}$	$10.5 - sp_0 = -1.65\sqrt{sp_0(1 - p_0)}$	$\underline{s}$
0.05	190	96	344	344
0.1	101	34	169	169
0.15	71	19	112	112
0.2	57	12	82	82
0.25	49	9	66	66
0.3	43	7	54	54
0.35	40	6	46	46
0.4	38	5	39	39
0.45	37	4	34	37
0.5	36	3	31	36

note, however, that the described procedure does not guarantee that correspondent sampling centroids belong to the same cluster in the "true" dataset partition. If this is not the case the suggested "arithmetic mean" formula becomes questionable. One way to overcome this difficulty is to use the maximal matching procedure mentioned in Sect. 5. In our case we have to employ this methodology for a comparison between two single samples. We recall that a sample is constructed by the union of several "elite" samples. Since in this section we do not work with samples' union, the procedure may fail to generate good results.

In what follows we introduce conditions under which the "arithmetic mean" formula is justified in the framework of the GMM. The results are provided at a significance level α and we denote by $Z_{1-\alpha/2}$ the $1 - \alpha/2$ quantile of the standard normal distribution. Consider now sample S_1 and denote the arithmetic mean of the cluster $S_1 \cap \pi_i$ by μ_i^1. We estimate the distance from the "true" cluster π_i center μ_i and the sample's mean μ_i^1. From normality of the probability distribution of each cluster one has

$$\left\| \mu_i - \mu_i^1 \right\| \leq Z_{1-\alpha/2}\sqrt{\frac{n}{s_0}}\sigma. \tag{31}$$

We wish to choose the sample size so that

$$\left\| \mu_i^1 - \mu_i \right\| < \left\| \mu_i^1 - \mu_t \right\|, \ t \neq i \tag{32}$$

(see Fig. 1). It is easy to see that this condition is satisfied if

$$4Z_{1-\alpha/2}\sqrt{\frac{n}{s_0}}\sigma < \left\| \mu_t - \mu_i \right\|. \tag{33}$$

This inequality leads to the lower bound for s_0 as follows:

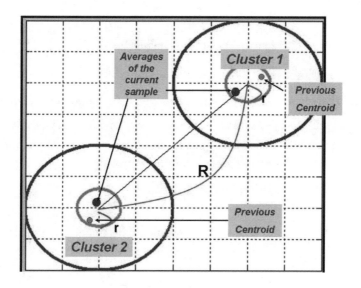

Fig. 1. Centroid separation, $r = 2Z_{1-\alpha/2}\sqrt{\frac{n}{s_0}}\sigma$ and $R = \|\mu_t - \mu_i\|$

$$\frac{16Z_{1-\alpha/2}^2 n\sigma^2}{d_\mu^2} < s_0, \qquad (34)$$

where

$$d_\mu = \min_{t,i}(\|\mu_t - \mu_i\|, \ t \neq i). \qquad (35)$$

Substituting s_0 by its average value sp_0 we obtain:

$$\frac{16Z_{1-\alpha/2}^2 n\sigma^2}{p_0 d_\mu^2} < s. \qquad (36)$$

Finally we denote

$$\tau = \frac{d_\mu}{\sigma} \qquad (37)$$

and simplify this so that the lower bound for the sample size s becomes

$$\frac{16Z_{1-\alpha/2}^2 n}{p_0 \tau^2} < s. \qquad (38)$$

This yields (32). Note that the ratio τ is unknown. In Sect. 6.3 we estimate this ratio using a resampling procedure. The estimate for τ (see (51)) completes the section.

6.3 Ratio Estimation

Consider several samples $S_j, j = 1, 2, ...$ from the dataset X and denote

$$\pi_t^{(j)} = S_j \cap \pi_t, \ t = 1, \ldots, k. \tag{39}$$

We write the "final" centroids of these clusters as

$$c_t^{(j)}(f) = \mu_t + \varepsilon_t, \quad t = 1, \ldots, k, \tag{40}$$

where ε_t are random vectors with normally distributed independent coordinates having mean zero and standard deviations $\sigma\sqrt{1/\left|\pi_t^{(j)}\right|}$. We denote $\mu_{t_1} - \mu_{t_2} + \varepsilon_{t_1} - \varepsilon_{t_2}$ by L_{t_1,t_2} and provide the following expression for the squared distance between two final sample centroids:

$$D_{t_1,t_2}^{(j)} = \left\| c_{t_1}^{(j)}(f) - c_{t_2}^{(j)}(f) \right\|^2 = \left\| L_{t_1,t_2} \right\|^2. \tag{41}$$

Generally, the random values $D_{t_1,t_2}^{(j)}$ are not independent. For instance, the covariance matrix of L_{t_1,t_2} and L_{t_2,t_3} is

$$cov\left(L_{t_1,t_2}, L_{t_2,t_3}\right) = \frac{\sigma^2}{\left|\pi_{t_2}^{(j)}\right|} I \tag{42}$$

where I is the $n \times n$ identity matrix. However, as sample size grows the correlation tends to 0. We disregard this dependence and write

$$D_{t_1,t_2}^{(j)} = (\sigma_{t_1,t_2}^{(j)})^2 \left\| \frac{\mu_{t_1} - \mu_{t_2}}{\sigma_{t_1,t_2}^{(j)}} + \varepsilon_{t_1,t_2} \right\|, \tag{43}$$

where ε_{t_1,t_2} are standard normally distributed vectors with independent coordinates and

$$\sigma_{t_1,t_2}^{(j)} = \sigma \sqrt{\frac{1}{\left|\pi_{t_1}^{(j)}\right|} + \frac{1}{\left|\pi_{t_2}^{(j)}\right|}}. \tag{44}$$

Due to the model assumptions of equal standard deviations for all the vector coordinates, an efficient accurate estimate for σ can be obtained through the averaging

$$\tilde{\sigma} = \frac{1}{n} \sum_{x \in X} \sigma(x_i) \tag{45}$$

where $\sigma(x_i)$ is the sample standard deviation of the ith coordinate of vectors $x \in S_j$. We replace σ by $\tilde{\sigma}$ and consider the worst-case scenario of equal size clusters

$$D_{t_1,t_2}^{(j)} = \sigma_\nu^2 \left\| \frac{\mu_{t_1} - \mu_{t_2}}{\tilde{\sigma}_\nu} + \varepsilon \right\|^2$$

standard normal distributed coordinates, ν is a cluster size, and

$$\sigma_\nu^2 = \frac{2\sigma^2}{\nu}.$$

Hence

$$\tau_{t_1,t_2}^{(j)} = \frac{D_{t_1,t_2}^{(j)}}{\sigma_\nu^2}$$

has a noncentral chi-square distribution and one has to estimate the value

$$\tau_0^{(j)} = \min_{t_1 < t_2} (\tau_{t_1,t_2}^{(j)}).$$

Note that the distribution of $D_{t_1,t_2}^{(j)}$ is independent of j. This is commonly the case for sampling with replacement. For n independent standard normally distributed random variables $Z_1, ..., Z_n$ and constants $\delta_1, ..., \delta_n$ the random variable $d = \sum_{j=1}^{n} (Z_j + \delta_j)^2$ has a noncentral chi-squared distribution with n degrees of freedom and noncentrality parameter

$$\lambda = \sum_{j=1}^{n} \delta_j^2 \qquad (46)$$

(see, for example, [250]). If $\lambda = 0$ the distribution is the well-known regular chi-squared distribution. If $\lambda \neq 0$ it is the noncentrally chi-squared distribution. The mean of this distribution is given by

$$\mu = n + \lambda. \qquad (47)$$

We note that there is no standard definition of the noncentrality parameter in the literature. Some authors denote the parameter as λ, while others divide the right-hand side in (46) by 2. There are a number of different representations for the density function of the chi-squared noncentral distribution by means of various series expansions. An alternative approach is to use the appropriate F ratio instead of separate estimation of the parameter σ. This leads to a more complicated model of F noncentral distribution and will not be pursued in this chapter.

In the present case the noncentrality parameter of $\tau_{t_1,t_2}^{(j)}$ is

$$\lambda_{t_1,t_2}^{(s)} = \frac{\|\mu_{t_1} - \mu_{t_2}\|^2}{\sigma_\nu^2}. \qquad (48)$$

If we denote by $\chi^2 (x, \lambda_{t_1,t_2}^{(s)}, n)$ the cumulative distribution function (cdf) of $\tau_{t_1,t_2}^{(s)}$, then the cdf of $\tau_0^{(s)}$ is given by

$$F(x) = 1 - \prod_{t_1 < t_2} (1 - \chi^2(x, \lambda_{t_1,t_2}^{(s)}, n)). \qquad (49)$$

Estimation of the parameters in this expression is a very difficult problem due to the complicated form of $F(x)$. However, we use a normal approximation of $\chi^2 (x, \lambda_{t_1,t_2}^{(s)}, n)$ for large values of the noncentrality parameter λ. As numerical experiments show for $\lambda > \lambda_0 = 100$ the cdf of $\chi^2 (x, \lambda_{t_1,t_2}^{(s)}, n)$ is well approximated by the normal distribution and $F(x)$ is well approximated by $\chi^2(x, \lambda_0^{(s)}, n)$. Hence we assume that

$$F(x) \approx \chi^2\left(x, \lambda_0^{(s)}, n\right), \quad \text{where } \lambda_0^{(s)} = \min_{t_1 < t_2} \lambda_{t_1, t_2}^{(s)}.$$

Let

$$(t_1^m, t_2^m) = \arg\min_{t_1 < t_2} \lambda_{t_1, t_2}^{(s)}, \tag{50}$$

that is $\mu_{t_1^m}$ and $\mu_{t_2^m}$ are the nearest centroids of $\Pi = \{\pi_1, \ldots, \pi_k\}$. We use an appropriate selection of the sample size to approximate the cdf $\chi^2(x, \lambda_0^{(s)}, n)$ by the normal distribution. The algorithm suggested here is based on (47). For a user-supplied positive tolerance tol do the following:

1. Set counter $t = 1$.
2. Set $j = 1$.
3. Draw random sample S_j of size s (if $t = 1$, then pick s from Table 1).
4. Use the k-means clustering algorithm to partition S_j into k clusters.
5. Compute $\tilde{\sigma}$.
6. Compute $\tau_0^{(j)}$, using centroids of S_j instead of μ_{t_i} to evaluate $D_{t_1, t_2}^{(j)}$.
7. Average $\tau_0^{(j)}$ over M samples $S_1, \ldots, S_M$.
8. *If* the stopping criterion $\left|\tau_0^{(j)} - \tau_0^{(j-1)}\right| < \text{tol} * \tau_0^{(j)}$ is met
   then if $[\tau_0^{(j)} > \sqrt{\lambda_0}]$
        then stop and accept the resulting parameter τ_0
        else set $s = 2s$, $t = t + 1$ and goto step 2.
   else set $j = j + 1$ and goto step 3.

We replace the average μ by τ_0 in (47), λ by its value from (48) with (t_1^m, t_2^m), and σ_ν^2 by its expression and obtain

$$\frac{sp_0}{2}(\tau_0 - n) = \frac{\|\mu_{t_1} - \mu_{t_2}\|^2}{\sigma^2}.$$

This yields the following estimate for τ

$$\tau = \sqrt{\frac{(\tau_0 - n)}{2} sp_0}. \tag{51}$$

One can also employ an extended procedure consisting of parallel estimation of the ratio and construction of initial centroids. That is, one can execute the averaging of cluster centroids with an appropriate choice of the initial centroids at every iteration as described above. We shall not pursue this approach in this chapter.

The initial centroids for the k-means like clustering algorithm applied to the entire dataset are the average of the corresponding final centroids generated for all M samples as described in Sect. 6.2.

7 Numerical Experiments

In order to evaluate the algorithms described in Sect. 5, we quantify how well the clustering results correspond to the "true" data structure. Dhillon and Modha [136] have used the spherical k-means algorithm for clustering text data. In one of the experiments of [136] the algorithm was applied to a data set containing 3,893 documents. This dataset has been considered in several works, (see for example [278, 279, 281, 432]). This dataset is known to be well separated. (The dataset is available at ftp://ftp.cs.cornell.edu/pub/smart)

- DC0–Medlars Collection (1,033 medical abstracts);
- DC1–CISI Collection (1,460 information science abstracts);
- DC2–Cranfield Collection (1,400 aerodynamics abstracts).

We selected 600 "best" terms (see [133] for term selection details), and calculated the 30 leading principal components. Note that each component covers no more than 5% of the total dispersion. Therefore we compared two projection pursuit indexes of Friedman, and Intrator and Cooper (see Sect. 3). Multimodality of the components' distributions can be observed only for the first two components. This fact is reflected by the behavior of the Intrator and Cooper index. The Friedman index is more sensitive to departure from normality in the tails of the distribution. The fifth component possesses the heaviest tail. Figures 2 and 3 present graphs of the two indexes.

However, the Intrator and Cooper index allocates precisely the two leading components as the informative features. Partitions into three clusters using

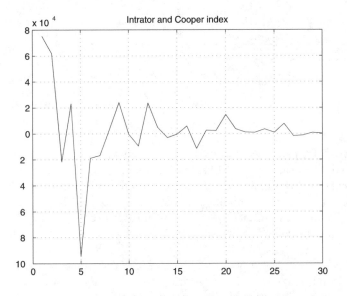

Fig. 2. The Intrator and Cooper index

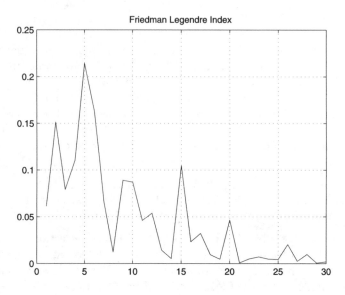

Fig. 3. The Friedman index

Table 2. Sample size is 40. Number of samples is 30

Number of components	Method	Misclassified items	Time (s)	Iterations	Cramer	Rand
2	SCA	127	4.14	2	0.950	0.958
2	Optimization	152	4.78	2	0.940	0.951
2	CE	732	19.59	6	0.731	0.816
5	SCA	118	5.98	3	0.953	0.961
5	Optimization	122	5.89	2	0.952	0.960
5	CE	520	60.78	18	0.813	0.839
10	SCA	125	21.02	33	0.951	0.959
10	Optimization	108	190.11	45	0.957	0.964
10	CE	637	230.64	67	0.779	0.815

the "naïve approach" were built by the described algorithms. The results are compared by means of misclassification quantities, the Cramer correlation coefficient, and the clustering matching Rand's coefficient, see Sect. 4.2. In addition, we consider a modification of the Sampling Clustering Algorithm, which consists of applying a search optimizing procedure for the objective function $\hat{R}$ constructed on the sample $\hat{S}_t$ (we suggest $\rho = 0.25$). Tables 2 and 3 summarize outcomes obtained for two "extreme" cases that samples with replacement cover about 30% and 150% of the data, respectively. We conclude that the Sampling Clustering Algorithm seems to be more robust and reliable for each choice of the principal components presented.

Table 3. Sample size is 80. Number of samples is 80

Number of components	Method	Misclassified items	Time (s)	Iterations	Cramer	Rand
2	SCA	140	5.55	2	0.945	0.954
2	Optimization	152	15.78	2	0.940	0.951
2	CE	236	49.67	7	0.912	0.926
5	SCA	94	8.23	2	0.963	0.969
5	Optimization	123	32.58	2	0.951	0.960
5	CE	396	116.31	18	0.853	0.880
10	SCA	103	18.92	3	0.959	0.966
10	Optimization	99	42.78	2	0.961	0.967
10	CE	831	202.13	29	0.781	0.781

Acknowledgments

The research of Kogan and Nicholas was supported in part by the Fulbright Program and Northrop Grumman Mission Systems. Kogan's work was also supported by the United States–Israel Binational Science Foundation (BSF).

TMG: A MATLAB Toolbox for Generating Term-Document Matrices from Text Collections

D. Zeimpekis and E. Gallopoulos

Summary. A wide range of computational kernels in data mining and information retrieval from text collections involve techniques from linear algebra. These kernels typically operate on data that are presented in the form of large sparse term-document matrices (tdm). We present TMG, a research and teaching toolbox for the generation of sparse tdms from text collections and for the incremental modification of these tdms by means of additions or deletions. The toolbox is written entirely in MATLAB, a popular problem-solving environment that is powerful in computational linear algebra, in order to streamline document preprocessing and prototyping of algorithms for information retrieval. Several design issues that concern the use of MATLAB sparse infrastructure and data structures are addressed. We illustrate the use of the tool in numerical explorations of the effect of stemming and different term-weighting policies on the performance of querying and clustering tasks.

1 Introduction

Much of the knowledge available today is stored as text. It is not surprising, therefore, that data mining (DM) and information retrieval (IR) from text collections (text mining) has become an active and exciting research area; see for example [399]. As the vector space model (VSM) and matrix and vector representations are routinely used in DM and IR, it turns out that several performance critical kernels in these areas originate from computational linear algebra (CLA). Consider, for example, two typical operations: clustering and querying in the context of the VSM. Algorithms that implement them rely on modules standing at various levels in the hierarchy of linear algebra computations, from inner products to eigenvalue and singular value decompositions (SVD). As a result, the fields of DM and IR have been providing the ground for synergistic efforts between application specialists and researchers in CLA. The latter researchers understand the intricacies of designing effective matrix computations on modern computer system platforms, commonly used in DM and IR and contribute the design of performance critical kernels for DM and IR algorithms; see, for example [61, 63, 65, 76, 136, 272, 275].

This chapter presents TMG, a toolbox that helps the user in the two major phases of the VSM: the preprocessing, "indexing" phase, in which the index of terms is built, and the "search" phase, during which the index is used in the course of queries and other operations. In particular, TMG preprocesses documents to construct an index in the form of a sparse "term-document matrix," hereafter abbreviated by "tdm," and preprocesses user queries so as to make them ready for the application of an IR model. TMG is specifically oriented to the application of vector space techniques (see, e.g., [124, 310, 456]) that model documents as term vectors so that many IR tasks can be cast in terms of CLA. We will use the convention that $m \times n$ matrices represent tdms of n documents over an index of m terms. In view of the significant presence of CLA kernels in vector space techniques for IR, we felt that there was a "market need" for a MATLAB-based tdm generation system, as MATLAB is a highly popular problem-solving environment for CLA that enables the rapid prototyping of novel IR algorithms [7]. Therefore, TMG is written entirely in MATLAB and runs on any computer system that supports that environment. Even though MATLAB started as a "Matrix Laboratory," it is now equipped with a large number of facilities including data structures, functions, visualization, and interface building tools that make possible the rapid synthesis of entire suites of special purpose algorithms. Furthermore, it claims a very large user base that continuously contributes new software (such as TMG) on the Web.[1] See [191, 205], for example, for toolboxes that specialize in operations related to IR, e.g., algorithms based on spectral analysis of sparse matrices encapsulating graph structures.

TMG parses single files or entire directories of multiple files containing text, performs the necessary preprocessing, and constructs a tdm according to parameters set by the user. It is also able to renew existing tdms by performing efficient updates or downdates corresponding to the incorporation of new documents or deletion of existing documents.

We must emphasize that two critical components for TMG's operations are MATLAB's sparse matrix infrastructure and visualization tools. TMG can be used to complement algorithms and tools that work with tdms, e.g., [22, 61, 63, 76, 136, 289]. For example, TMG was used in recent experiments, including in [286, 364, 452, 453], and has already been requested by many researchers. We also expect TMG to be useful in instructional settings, by helping to create motivating examples in CLA and IR courses.

TMG is not tied to specific algorithms of the vector space model but includes, for convenience, MATLAB code for querying and clustering that can be used as template by users who want to perform such tasks with TMG. Interfacing TMG with codes in other languages is quite straightforward as there are several utilities for converting the objects in the underlying MATLAB storage class to other formats. The CLUTO clustering toolkit [257], for example,

---

[1]See, for example, the Link Exchange Center, www.mathworks.com/matlab central/fileexchange

inputs ASCII files containing the compressed sparse row (CSR) representation of matrices that can be obtained from TMG, while SDDPACK [289] provides MATLAB routines for converting to and from the Matrix Market format [74]. TMG was designed to provide several term-weighting options to the user [60,290] as well as the possibility of stemming. In addition to describing the design of the tool, we also report herein on the application of TMG as a preprocessor for IR tasks and its combination with a variety of term-weighting functions and stemming.

This chapter is organized as follows. In the rest of this section we briefly review some related efforts. Section 2 presents TMG, describing all its core functions, including the graphical user interface (GUI) and analyzing the various options that are provided in the package. Section 3 describes implementation issues, in particular the utilization of MATLAB's sparse matrix technology. Section 4 demonstrates the use of TMG on a public dataset, called BIBBENCH, and compares the performance of some query answering and clustering algorithms based on vector space models using various term-weighting schemes and stemming for data from the MEDLINE, CRANFIELD, CISI[2] and REUTERS-21578[3] collections. Section 5 provides concluding remarks. All numerical experiments were conducted on a 3 GHz Pentium 4 PC with 512 MB RAM running Windows XP and MATLAB 7.0. Runtimes were measured using MATLAB infrastructure for performance analysis, specifically `profile` and timing functions `tic, toc`.

1.1 Related Work

There exist already several tools for constructing tdms since IR systems that are based on vector space techniques (e.g., Latent Semantic Indexing, hereafter abbreviated as LSI) typically operate on rows and columns of such matrices; see, e.g., [5,310,379]. The *Telcordia LSI Engine*, for example, is a production-level IR architecture that contains components for generating sparse tdms [9,106] from text collections. *Lemur* [6] is a popular language modeling and IR toolkit written in C++. A recent powerful system that we have found to be particularly effective is the *General Text Parser* (GTP) [4,192]. GTP is a complete IR package written in C++ and Java and employing LSI; we used it to evaluate the results obtained from TMG. PGTP is an MPI-based parallel version of GTP. Other tools that one can find in the open literature are DOC2MAT [3], written in `perl` and developed in the context of the CLUTO IR package [257]; MC [8], written in C++ [129,136]; and the Unix shell script utility `countallwords` included in the PDDP package [77]. The above tools are implemented in high level or scripting languages (e.g., `C`, `C++`, `Java`, `Perl`). It is fair to say at the outset and will become clear from our description that TMG's current design is best suited for datasets of moderate size. For very large datasets, one would be better served by systems such as GTP.

---

[2] Available from ftp://ftp.cs.cornell.edu/pub/smart/
[3] Available from http://kdd.ics.uci.edu/

2 The Text to Matrix Generator

2.1 Outline

TMG is constructed to perform preprocessing and filtering steps that are typically performed in textual IR [40] (the names of the relevant MATLAB m-functions are in parentheses):

- Creation of the tdm corresponding to a set of documents (tmg);
- Creation of query vectors from user input (tmp_query);
- Updation of existing tdm by incorporation of new documents (tdm_update);
- downdation of existing tdm by deletion of specified documents (tdm_downdate).

The document preprocessing steps encoded by TMG are the following: (*i*) lexical analysis; (*ii*) stopword elimination; (*iii*) stemming; (*iv*) index-term selection; (*v*) index construction. These steps are tabulated in Table 1.

Each element, α_{ij}, of a tdm can be expressed as

$$\alpha_{ij} = l_{ij}g_i n_{ij}, \tag{1}$$

where l_{ij} is a local factor that measures the importance of term i in document j, g_i is a global factor that measures the importance of term i in the entire collection, and n_{ij} is a normalization factor [380]. This latter is used to moderate bias toward longer documents [398]. The local, global term weighting and normalization options available in TMG are listed in Table 2. Symbol f_{ij} denotes term frequency, i.e., the number of times term i appears in document j; also,

$$p_{ij} = \frac{f_{ij}}{\sum_k f_{ik}}, \quad \text{and} \ \ b(f_{ij}) = \begin{cases} 1, \text{ if } f_{ij} \neq 0 \\ 0, \text{ if } f_{ij} = 0. \end{cases}$$

Table 1. Steps in function tmg

Function tmg
Input: filename, OPTIONS
Output: tdm, dictionary, and several optional outputs;
parse files or input directory;
read the stoplist;
for each input file,
parse the file (construct dictionary);
end
normalize the dictionary (remove stopwords and too long or too short terms, stemming);
construct tdm;
remove terms as per frequency parameters;
compute global weights;
apply local weighting function;
form final tdm;

Table 2. Term-weighting and normalization schemes [60, 110, 290, 380]

Symbol Name		Type
Local term-weighting (l_{ij})		
t	Term frequency	f_{ij}
b	Binary	$b(f_{ij})$
l	Logarithmic	$\log_2(1 + f_{ij})$
a	Alternate log [290]	$b(f_{ij})(1 + \log_2 f_{ij})$
n	Augmented normalized term frequency	$(b(f_{ij}) + (f_{ij}/\max_k f_{kj}))/2$
Global term-weighting (g_i)		
x	None	1
e	Entropy	$1 + (\sum_j (p_{ij} \log_2(p_{ij}))/\log_2 n)$
f	Inverse document frequency (IDF)	$\log_2(n/\sum_j b(f_{ij}))$
g	Gfldf	$(\sum_j f_{ij})/(\sum_j b(f_{ij}))$
n	Normal	$1/\sqrt{\sum_j f_{ij}^2}$
p	Probabilistic inverse	$\log_2((n - \sum_j b(f_{ij}))/\sum_j b(f_{ij}))$
Normalization factor (n_{ij})		
x	None	1
c	Cosine	$(\sum_j (g_i l_{ij})^2)^{-1/2}$

It must be noted that TMG does not restrict the separating delimiter to be an end-of-file character; hence the number of documents corresponding to the collection would be at least as large as the actual number of (valid) files processed by TMG.

2.2 User Interface

The user interacts with TMG by means of any of the aforementioned MAT-LAB functions or via a GUI, implemented as function tmg_gui. The GUI facilitates user selection of the appropriate options among the many alternatives available at the command-line level. A user who desires to construct a tdm from text will either use tmg or tmg_gui. The specific invocation of the former is of the form:

outargs=tmg('fname', OPTIONS);

where outargs stands for the output list:

[A, dictionary, global_wts, norml_factors, words_per_doc, titles, files, update_struct].

The tdm is stored as a MATLAB sparse double array A, while dictionary is a char array containing the collection's distinct words, and update_struct contains the essential information for the collection's renewal (see Sect. 3.3).

Table 3. TMG outputs

A	resulting tdm;
dictionary	collection's dictionary (char array);
global_wts	vector of global weights
norml_factors	vector of document norms prior to normalization;
words_per_doc	vector containing statistics for each document;
titles	titles of each document (cell array);
files	processed filenames with set title and document's first line (cell array);
update_struct	structure containing necessary data for renewal;

The other output arguments store statistics for the collection. The full list of output arguments is tabulated in Table 3.

Argument fname specifies the individual file(s) to be processed or the directory name that contains them. In the latter case, TMG recursively processes included subdirectories and files. It is assumed that all files contain valid data. In particular, files are assumed to contain plain ASCII text or that a special filter that can convert them to such a format is available and properly linked to TMG. Currently, TMG can process Adobe Acrobat PDF and POSTSCRIPT documents provided Ghostscript's ps2ascii utility is available. Filenames suffixed with html or htm are assumed to be ASCII files with html markups; TMG processes them by stripping the corresponding tags using the strip_html function.

The options available at the command line to the user of tmg are set via the fields of the MATLAB OPTIONS structure tabulated in Table 4. Field delimiter specifies the delimiter that separates individual documents within the same file. The default delimiter is a blank line, in which case TMG is likely to generate more "documents" than the number of files given as input. Field line_delimiter specifies if the delimiter takes a whole line of text. Field stoplist specifies the file containing the stopwords, i.e., the terms excluded from the collection's dictionary [60]. The current release of TMG contains a stoplist obtained from GTP [4]. Field stemming indicates whether stemming is to be used; this is performed by stemmer, our MATLAB implementation of a modified version of Porter's algorithm [361, 362]. stemmer can also be called directly from the command line. To validate our implementation, we compared our results and verified that they coincided with the word list and corresponding stems listed in [361].

In the current version of TMG, stopword removal takes place before stemming. Therefore, care is required in adding terms in the stoplist, as we might need to provide their variants as well. In particular, if in a bibliography file we wish to dispose of the words "author" and "authors" we would need to add both to the stoplist. It is easy to alter this in TMG so as to apply stemming on the stoplist as well as on the dictionary. One disadvantage is that this could lead to the removal of terms that share the same stem with a stop-

Table 4. OPTIONS fields

delimiter	String specifying the "end-of-document" marker for tmg. Possible values are emptyline (default), none_delimiter (treats each file as a single document) or any other string
line_delimiter	Variable specifying if the delimiter takes a whole line of text (default, 1)
stoplist	Filename for stopwords (default no name, meaning no stopword removal)
stemming	A flag that indicates if stemming is to be applied (1) or not (0) (default stemming=0)
min_length	Minimum term length (default 3)
max_length	Maximum term length (default 30)
min_local_freq	Minimum term local frequency (default 1)
max_local_freq	Maximum term local frequency (default Inf)
min_global_freq	Minimum number of documents for a term to appear to insert it in the dictionary (default 1)
max_global_freq	Maximum number of documents for a term to appear to insert it in the dictionary (default Inf)
local_weight	Local term weighting function (default 't', possible values 't', 'b', 'l', 'a', 'n')
global_weight	Global term weighting function (default no global weighting used, 'x', possible values 'x', 'e', 'f', 'g', 'n', 'p')
normalization	Flag specifying if document vectors are to be normalized ('c') or not ('x') (default)
dsp	Flag specifying if results are to be printed in the command window (1, default) or not (other)

word. Another option that would be easy to incorporate in TMG is to use two stoplists: one containing basic stopwords and the other, a stoplist generator, for domain-specific terms that would be useful to preprocess by stemming and then use it as stoplist. Overall, given the current options in TMG, it is not difficult to enrich the current filtering steps to help process "dirty text," containing typos, adhoc abbreviations, special symbols, etc. [99].

Parameters min_length, max_length are thresholds used to exclude terms that are out of range; e.g., terms that are too short are likely to be of little value in indexing while very long ones are likely to be misprints. Parameters min_local_freq, max_local_freq, min_global_freq, and max_global_freq are also filtering parameters, thresholding based on frequency of occurrence. The last OPTIONS field, dsp, indicates if the intermediate results are printed on the MATLAB command window.

Function tmg_query uses the dictionary returned by tmg and constructs, using the same processing steps as TMG, a "term query" array whose columns are the (sparse) query vectors for the text collection. The function is invoked as follows:

```
[Q, wds_per_query, titles, files]=tmg_query('fname',dictionary, OPTIONS);
```

Here, OPTIONS contains fields that are a subset of those used in tmg; for details see the code documentation.

Graphical User Interface

As described thus far, the main toolbox functions tmg and tmg_query offer a large number of options. Moreover, it is anticipated that future releases will further increase this number to allow for additional flexibility in operations and filetypes handled by TMG. In view of this, a GUI we would be calling TMG_GUI, which is depicted in Fig. 1, was created to facilitate interaction. This is instantiated by means of function tmg_gui. The GUI design was facilitated by the interactive MATLAB tool GUIDE. TMG_GUI consists of two frames: one provides a set of four mutually exclusive radio buttons, corresponding to the basic functions of TMG, along with a set of radio buttons, edit boxes, lists, and toggle buttons for all required input arguments; the other provides a set of items for the optional arguments of tmg, tmg_query, and update routines. After specifying all necessary parameters, and the Continue button is clicked, TMG_GUI invokes the appropriate function. The progress of the program is shown on the screen; upon finishing the user is queried if and where he wants the results to be saved; results are saved in MATLAB-mat file(s), i.e., the file format used by MATLAB for saving and exchanging data.

3 Implementation Issues

We next address some issues that relate to design choices made regarding the algorithms and data structures used in the tool. Overall, TMG's efficiency is greatly aided by the use of MATLAB's sparse matrix infrastructure and an effective implementation of inverted indexes.

3.1 Sparse Matrix Technology

One important goal in the design of TMG was to employ data structures that would be efficient regarding: (*i*) the costs of creating and updating them, (*ii*) the overall storage requirements, and (*iii*) the processing of the kernel IR operations. Tdms are usually extremely sparse; e.g., see Table 10 that tabulates the statistics for some well-known collections used to benchmark IR algorithms, and Table 7 for the statistics for our BIBBENCH collection: Approximately 98% or more of the entries of the corresponding tdms are 0. Therefore, a natural object for representing tdms is the sparse matrix. Indeed,

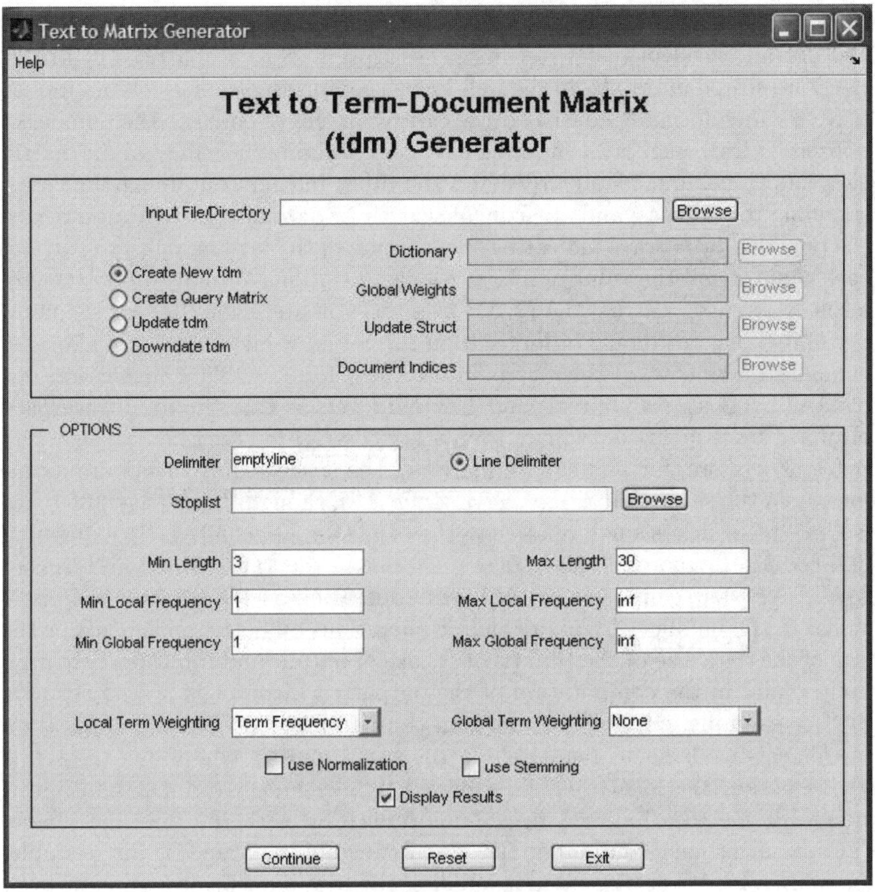

Fig. 1. TMG_GUI

with the current popularity of VSM-based techniques, sparse matrix representations have become popular in IR and are the subject of investigation; see e.g., [194, 238]. It is worth noting that recent studies suggest that sparse matrices are preferable for IR over other implementations, such as inverted indexes [194, 195]. Inverted indexes, for example, complicate the implementation of non-Boolean searches and dimensionality reduction transformations that are at the core of LSI [39]. Nonetheless, TMG employs an inverted index as an intermediate data structure to aid in the assembly of the sparse tdm.

After parsing the collection (cf. Sect. 3.2), cleaning and stemming the dictionary, each cell array for the posting list is copied to another, each element of which is a MATLAB sparse column vector of size n. This latter array is finally converted to the sparse tdm using function `cell2mat`.

MATLAB provides an effective environment for sparse computations built around the concept of a "sparse array," a special MATLAB class that economizes storage and operations by utilizing mature technology; see [189, 190] for an excellent, early, technical description. Sparse matrices in MAT-LAB are stored internally in the well-known compressed sparse column format (CSC), which formally consists of two arrays of length equal to the number of nonzero entries, nnz, one consisting of reals containing the values of the matrix elements in column major order and the other integer containing the corresponding row indices; and an array of size $n+1$ containing an integer index to the previous two arrays indicating the location of the leading nonzero entry of each column and the value of nnz at the last position. Actually, upon the creation of a sparse matrix, MATLAB uses an estimate, nzmax(A) for the number of its nonzeros (equal or larger than the actual value of nnz) and allocates enough storage to store the matrix in the above format [189]. Current versions of MATLAB use 8 byte reals and 4 byte integers so that the total workspace occupied by a sparse nonsquare tdm A is $Mem(A) = 12\text{nzmax}(A) + 4(n + 1)$ bytes. Therefore, for non square matrices, the space requirements are asymmetric, in the sense that $Mem(A) \neq Mem(A^\top)$, though the storage difference is $4|m-n|$, which is small relative to the total storage required. By expressing and coding the more intensive manipulations in the TMG toolbox in terms of MATLAB sparse operations, the cost of operating on tdms becomes proportional to the number of real arithmetic operations on nonzero elements or the size of the data size of the tdm (that is, size of output and input participating nontrivially in the computation of the output), whichever is larger. Formula (1), for example, implies that the tdm can be obtained from the application of element-by-element operations on the sparse matrix containing the terms f_{ij} to obtain the local weights, followed by element-by-element multiplication with the tdm followed by left multiplication with the diagonal matrix (in sparse format) containing the global weights g_i. Table 5, for example, shows MATLAB statements for building the tdm for scheme lnc. New term weighting formulas (e.g., [110]) can be easily programmed in the system.

It is worth noting here that had we opted to build the target tdm directly as a sparse matrix in the course of the reading phase, it would have necessitated fast updates (creating new rows and columns, changing individual elements), which would have been inefficient, especially in the absence of a good a priori estimate of the matrix size and total number of nonzeros.

As already mentioned, sparse representations are employed by other systems as well. GTP and the Telcordia LSI Engine systems, for example, use the Harwell–Boeing format [4, 106, 192], while the MC toolkit [8] also uses

Table 5. MATLAB commands to build the tdm A for scheme lnc from the frequency table $F = (f_{ij})$

```
[i, j, L]= find(F); L=log2(L+1);
A=sparse(i, j, L, size(F,1),size(F, 2))
A = spdiags(1./sqrt(sum(F.^2,2)), 0, size(F,1), size(F,1))*A;
A = A*spdiags(1./sqrt(sum(A.^2,1))', 0, size(A,2), size(A,2));
```

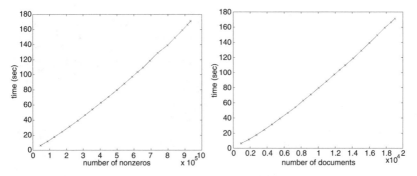

Fig. 2. TMG runtimes (s) vs. the number of nonzeros in the tdm (left); vs. the number of documents (right)

the CSC format. The authors of [194] use the CSR format to store instead "document-term" matrices; this, of course, is equivalent to our approach. On the other hand, the experiments in [238] assume a CSR representation for term-document matrices.

We next experimentally illustrate the dependence of TMG's runtime on aspects of the dataset size. In this as well as in Sect. 4, we experimented with datasets created from the REUTERS-21578 collection. We kept only those texts that contained nonempty text bodies and called the resulting set, consisting of 19,042 documents, REUT-ALL. We then organized the collection in 22 files, which we labeled REUTi, where $i = 1, \ldots, 22$. In the sequel, we would be using the notation REUT$[i : j]$ to denote the dataset consisting of files REUTi up to and including REUTj.

Figure 2 shows the runtimes of TMG to build tdms from each of the 22 file collections REUT1, REUT$[1 : 2]$, up to REUT$[1 : 22]$, vs. the number of nonzero elements and the number of documents. The figure suggests that the time taken by TMG depends linearly on the number of nonzeros of the tdm. The dependence also appears to be linear in the number of documents. We also illustrate the performance of two kernel CLA operations for IR, specifically matrix–vector multiplication and the computation of the largest singular value and corresponding singular vectors using the native MATLAB function for sparse SVD (`svds`); the latter is based on the implicitly restarted Arnoldi method [308]. Results are shown in Fig. 3.

3.2 Dictionary Construction

Central to the operation of TMG are the steps of document parsing and dictionary construction. TMG reads each document using function `strread`. This returns the tokens present in its input char array in a cell array of chars. All distinct terms present in the document are then obtained in a sorted order via function `unique`. At the same time, the procedure creates a (local) posting list for these terms, that is, pairs containing the number of occurrences and the

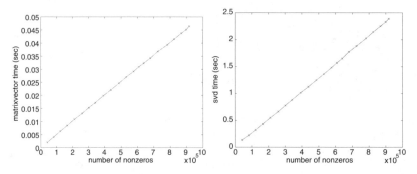

Fig. 3. Runtimes (s) of matrix–vector multiplication (left) and SVD (right) where the matrix is the tdm constructed by TMG vs. the number of nonzeros in the tdm

document identifier for each term. Assuming that we keep a "running inverted index" of all documents processed up to step $i - 1$, we can apply the procedure iteratively as follows: at steps $i = 2, \ldots$, we first create the local term vector and posting list and then use it to update the running inverted index. One weakness of this approach is that it requires as many calls to functions `unique` and `ismember` as there are documents, something that we found to be very time consuming. Another approach would be to proceed by appending to the dictionary's cell array the new terms in the document and keeping track of the document indices containing each word. This would necessitate only one call to `unique` to form the inverted index but with high cost in memory, since we would need to store first all tokens in the collection. The memory penalty is further accentuated by the fact that `cell` data structures have a higher memory overhead than sparse numeric arrays. Based on the above observations, we designed a simple but effective scheme to construct the inverted index. In particular, we still use a running inverted index but update it using a block inverted index consisting of data from N documents at a time. The functions that implement the above operations are called `unique_words` and `merge_dictionary`, while we use `update_step` to designate the block size N. Selecting $N = n$ or $N = 1$, the above approach reduces to the first and second of the aforementioned methods. Figure 4 depicts the runtimes of TMG for the REUT-ALL collection, for $N = [10, 20, 50, 100, 200, 500, 1000 : 1000 : 10000]$. As we see, TMG's performance peaks at intermediate values of N. Finally, although it is not clear from this figure, larger values of N can further increase runtime because of disk accesses. The effects of this blocked approach to building the dictionary together with an incremental approach to constructing the tdm, presented in Sect. 3.3, are the subject of our current investigations toward better tuning of TMG's performance. Overall, however, because TMG does not currently implement text or index compression (see e.g., [40, 444]) it is better suited for datasets of moderate size.

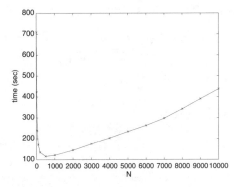

Fig. 4. TMG runtimes (s) over the **update_step** parameter N

3.3 TMG **for Document Renewal**

The efficient updating or downdating tdms is of importance as it is key to the maintenance of document collections. It can also lead to important CLA issues related to the design of effective algorithms for fast SVD updates; see for example [310, 445, 456]. In order to retain independence from the underlying VSM, we are concerned here with simple tdm updates that result in a matrix that is identical with the one that would have been created were all documents available from the beginning and TMG applied to all. In other words, we designed updating operations that maintain the integrity of the resulting tdm. To this end, TMG includes functions **tdm_update** and **tdm_downdate** for modifying an existing tdm so as to take into account document arrival and/or deletion. Any document arrival or deletion is likely to change the size of the tdm as well as specific entries: nontrivial document arrival will certainly change the number of tdm columns and the number and labeling of the rows, because terms satisfying the filtering requirements are encountered or removed; and/or terms in the original dictionary that were excluded by filtering become valid entries. Hence, in order to update correctly, the entire dictionary of terms encountered during parsing (before filtering) must be maintained together with the corresponding term frequencies. This information is also sufficient for the proper update of the tdm when parameters such as the maximum and/or minimum global frequencies change. Therefore, as long as updates are anticipated, when TMG is run, the user must select to save these items (TMG_GUI prompts the user accordingly). TMG saves them in a MATLAB structure array, denoted by **update_struct**. We avoid using one full and one normalized (postfiltering) dictionary by working only with the full dictionary and a vector of indices indicating those terms active in the normalized dictionary, stored in the same structure array. For **tdm_downdate**, the user specifies the relevant **update_struct** and a vector of integer indices identifying the documents to remove. We evaluate the performance of renewal in Sect. 4.2.

Table 6. GTP and TMG runtimes (in seconds)

Toolbox-Collection	REUT-ALL	MEDLINE	CRANFIELD	CISI
TMG	169.52	8.27	10.05	14.28
GTP	96.84	4.52	8.2	4.96

4 Experimental Results

To check the results obtained from TMG, we first used it to build the tdms from the MEDLINE, CRANFIELD, and CISI collections and confirmed that they were the same as those obtained using GTP, except for expected differences due to the fact that the two packages follow contrasting approaches to handle terms containing digits: TMG (resp. GTP) excludes (resp. includes) terms that are solely composed of numeric characters but keeps (resp. drops) words combining letters and numeric characters. We also show, in Table 6, the runtimes of TMG and GTP for the aforementioned datasets as well as for the set REUT-ALL, described in Sect. 3. Results with GTP were obtained on a system running Linux with the GCC 2.95 compiler. In view of the fact that TMG consists of MATLAB code, it is quite efficient, albeit slower than GTP. Furthermore, as mentioned earlier, GTP's lead is expected to increase for very large datasets.

4.1 The BIBBENCH Dataset

To illustrate the use of TMG we created a new dataset, called BIBBENCH, consisting of three source files from publicly accessible bibliographies[4] in BIBTEX (the bibliography format for LaTeX documents), with characteristics shown in Table 7. The first, we call BKN, is a 651-entry bibliography contained in this book, though loaded sometime before printing and therefore not corresponding exactly to the final edition. The major theme is, of course, clustering . The second bibliography, BEC is from http://jilawww.colorado.edu/bec/bib/, a repository of BIBTEX references from topics in atomic and condensed matter physics on the topic of Bose–Einstein condensation. When downloaded (Jan. 24, 2005) the bibliography contained 1,590 references. The last bibliography, GVL, was downloaded from http://www.netlib.org/bibnet/subjects/ and contains the full 861-item bibliography of the 2nd edition (1989) of a well-known treatise on matrix computations [198]. The file was edited to remove the first 350 lines of text that consisted of irrelevant header information. All files were stored in a directory named BibBench. It is worth noting that at first approximation, the articles in BEC could be thought as belonging in one cluster ("physics"), whereas those in BKN and GVL in another ("linear algebra and information retrieval").

---

[4]The bibliographies are directly accessible from the TMG web site.

Table 7. BIBBENCH dataset

Feature	BEC	BKN	GVL	BIBBENCH_txx_s
Documents	1,590	651	860	3,101
Terms (indexing)	1712	1159	720	3,154
Stemmed terms	372	389	221	964
Avg. terms/document	41	38	28	37
Avg. terms/document (indexing)	13	13	8.40	12
tdm nonzeros (%)	0.74	1.00	1.15	0.36

We first used TMG to assemble the aforementioned bibliographies using the terms weighting, no global weighting, no normalization, and stemming (txx_s) thus setting as nondefault OPTIONS

```
OPTIONS.delimiter='@'; OPTIONS.line_delimiter=0;
OPTIONS.stoplist='bibcommon_words'; OPTIONS.stemming=1;
OPTIONS.min_global_freq =2; OPTIONS.dsp= 0
```

Therefore, any word that appeared only once globally was eliminated (this had the effect of eliminating one document from GVL). The remaining packet had 3,101 bibliographical entries across three plain ASCII files: KNB.bib, BEC.bib, and GVL.bib. The stoplist file was selected to consist of the same terms found in [192] augmented by keywords utilized in BIBTEX and referring to items that are not useful for indexing, such as author, title, editor, year, abstract, keywords, etc.

The execution of the following commands

```
[A, dictionary, global_wts, norml_factors, words_per_doc,...
    titles, files, update_struct]=tmg('BibBench',OPTIONS);
```

has the following effect (compacting for some pretty-printing):

```
======================================================================
Applying TMG for file/directory C:\TMG\BibBench...
======================================================================
Results:
======================================================================
Number of documents = 3101
Number of terms = 3154
Avg number terms per document (before normalization) = 36.9987
Avg number of indexing terms per document = 11.9074
Sparsity = 0.362853%
Removed 302 stopwords...
Removed 964 terms using the stemming algorithm...
Removed 2210 numbers...
Removed 288 terms using the term-length thresholds...
Removed 6196 terms using the global thresholds...
Removed 0 elements using the local thresholds...
```

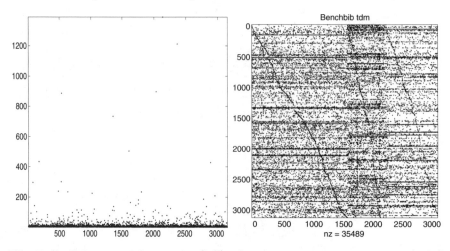

Fig. 5. BibBench term frequencies (left); tdm sparsity structure using tspy (right)

```
Removed 0 empty terms...
Removed 1 empty documents...
```

A simple combination of commands depicts the frequencies of the most frequently occurring terms. After running TMG as above, the commands

```
f=sum(A,2); plot(f,'.'); [F,I]=sort(f);
t=20; dictionary(I(end:-1:end-t+1),:)
```

plot the frequencies of each term (Fig. 5) and return the top $t = 20$ terms of highest frequency in the set, listed in decreasing order of occurrence below:

```
phy rev os condens instein lett atom trap comput algorithm
cluster method data ga usa system matrix linear matric mar
```

We next use TMG to modify the tdm so that it uses a different weighting scheme specifically **tnc** and stemming. This can be done economically with the **update_struct** computed earlier as follows:

```
update_struct.normalization='c'; update_struct.global_weight='n';
A=tdm_update([],update_struct);
```

Using MATLAB's **spy** command we visualize the sparsity structure of the tdm A in Fig. 7 (left). In the sequel we apply two MATLAB functions produced in-house, namely **pddp** and **block_diagonalize**. The former implements the PDDP(l) algorithm for clustering of term-document matrices [452]. We used $l = 1$ and partitioned in two clusters only, so that results are identical with the original PDDP algorithm [76]. In particular, classification of each document into one of the two clusters is performed on the basis of the sign of the corresponding element in the maximum right singular vector of matrix $A - Aee^\top/n$, where e is the vector of all 1s. The MATLAB commands and results are as follows:

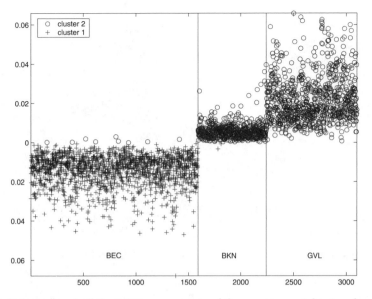

Fig. 6. Values of each of the 3,101 components of the maximum right singular vector v_{max} of the BIBBENCH dataset vs. their location in the set. The vertical lines that separate the three BIBTEXfiles and the labels were inserted manually

```
>> clusters = pddp(A,'svds','normal','1',2);
Running PDDP(1) for k=2...
Using svds method for the SVD...
Splitting node #2 with 3101 documents and 55.558 scatter value
    Leaf 3 with 1583 documents
    Leaf 4 with 1518 documents
Number of empty clusters = 0
PDDP(1) terminated with 2 clusters
```

Figure 6 plots the maximum singular vector v_{max} corresponding to the BIBBENCH dataset. Even though our goal here is not to evaluate clustering algorithms (there is plenty on this matter in other chapters of this volume!), it is worth noting that PDDP was quite good at revealing the two "natural clusters". Figure 6 shows that there are some documents from BEC (marked with '+') that were classified in the "clustering and matrix computations" cluster and very few documents from BKN and GVL that were classified in the "physics" cluster.

Finally, function `block_diagonalize` implements and plots the results from a simple heuristic for row reordering of the term-document matrix based on `pddp`. In particular, running

```
>> block_diagonalize(A, clusters);
```

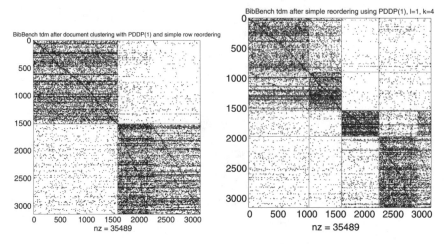

Fig. 7. Spy view of BIBBENCH tdm's for $k = 2$ (left) and $k = 4$ (right) clusters

we obtain Fig. 7 (right). This illustrates the improvement made by the clustering procedure. We note here that experiments of this nature, in the spirit of work described in [66], are expected to be useful for instruction and research, e.g., to visualize the effect of novel reordering schemes. Finally, Table 8, shows the size and 10 top most frequent terms (after stemming) for each of the four clusters obtained using PDDP(1). There were two "physics" clusters, the theme of another appears to be "linear algebra" while the theme of the last one is "data mining". The terms also reveal the need for better data cleaning [99], e.g., by normalizing or eliminating journal names, restoring terms, etc.: for instance, **numermath, siamnum** were generated because of nonstandard abbreviations of the journals *Numerische Mathematik* and *SIAM Journal of Numerical Analysis*. Terms **instein** and **os** were generated because of entries such as {E}instein and {B}os, where the brackets were used in the BIBTEX to avoid automatic conversion to lower case.

4.2 Performance Evaluation

Renewal Experiments

We next evaluate experimentally the performance of TMG when renewing existing tdms. We first ran TMG on the collection of 19,042 REUT-ALL documents and recorded the total runtime (169.52s) for tdm creation. We consider this to be one-pass tdm creation. We then separated the documents in $b = 2$ groups formed by REUT$[1 : j]$ and REUT$[(j + 1) : 22]$, $j = 1 : 21$, an "original group" of K documents and an "update group" with the remaining ones. We consider this to be tdm creation in $b = 2$ passes. We then ran TMG twice, first using **tmg** to create the tdm for the K documents and then **tdm_update**. We performed a similar experiment for downdating, removing in each step

Table 8. Ten most frequent terms for each of the four clusters of BIBBENCH using PDDP(1). In parentheses are the cluster sizes. We applied stemming but only minimal data cleaning

I (1,033)	II (553)	III (633)	IV (885)
phy	phy	matric	cluster
rev	instein	numermath	usa
os	rev	matrix	data
condens	condens	eigenvalu	comput
trap	os	siamnuman	mine
instein	lett	symmetr	algorithm
ga	ketterl	linalgapp	york
atom	atom	problem	analysi
lett	optic	linear	parallel
interact	mar	solut	siam

Table 9. Runtimes (s) for document renewal. To build the collection in one-pass took 205.89 s

K	tmg	up	Total	down	K	tmg	up	Total	down
925	6.00	271.70	277.70	0.11	10,963	88.88	82.52	171.39	0.34
2,761	18.00	224.05	242.05	0.13	11,893	97.92	70.55	168.47	0.38
3,687	24.75	203.24	227.99	0.16	12,529	104.64	63.63	168.27	0.44
4,584	31.83	185.09	216.92	0.19	13,185	110.67	56.14	166.81	0.39
5,508	39.70	167.25	206.95	0.17	14,109	119.92	46.77	166.69	0.45
6,429	47.19	151.63	198.81	0.20	15,056	128.92	37.58	166.50	0.48
7,319	54.69	137.28	191.97	0.22	15,996	139.86	27.16	167.02	0.52
8,236	63.22	121.78	185.00	0.25	16,903	149.66	19.23	168.89	0.56
9,140	71.47	108.53	180.00	0.31	17,805	159.58	11.66	171.24	0.61
10,051	80.39	93.91	174.30	0.31	18,582	166.61	6.78	173.39	0.66

the second part from the complete collection. Runtimes are summarized in Table 9. We observe that renewal is quite efficient, and in some cases approaches the one-pass creation. In any case, it clearly proves that renewing is much more efficient than recreating the tdm from scratch. Also, the gains from downdating (vs. rebuilding) are even larger. These experiments also suggest that for large datasets, even if the entire document collection is readily available and no further modifications are anticipated, it might be cost effective to build the tdm in multiple ($b \geq 2$) passes.

4.3 Evaluating Stemming and Term-Weighting

We next take advantage of the flexibility of TMG to evaluate the effect of different term weighting and normalization schemes and stemming in the context of query answering and clustering with VSM and LSI.

Query Answering

Our methodology is similar to that used by several other researchers in CLA methods for IR, see, for example, [290]. We experimented with all possible schemes available in TMG on standard data collections. In the case of LSI, we used as computational kernel the sparse SVD algorithm implemented by MATLAB's `svds` function. We note that this is just one of several alternative approaches for the kernel SVD in LSI (cf. [62, 64, 232, 304]) and that TMG facilitates setting up experiments seeking to evaluate their performance. A common metric for the effectiveness of IR models is the N-point interpolated average precision, defined by

$$p = \frac{1}{N} \sum_{i=0}^{N} \hat{p}(\frac{i}{N-1}), \tag{2}$$

where $\hat{p}(x) = \max\{p_i \mid n_i \geq xr, i = 1 : r\}$ is the "precision" at "recall" level $x, x \in [0, 1]$. Precision and recall after i documents have been examined are $p_i = n_i/i$, and $r_i = n_i/r$, respectively, where, for a given query, n_i is the number of relevant documents up to the i-th document, and r is the total number of relevant documents. We used this measure with $N = 11$ (a common choice in IR experiments) for three standard document collections: MEDLINE, CRANFIELD, and CISI whose features, as reported by TMG, are tabulated in Table 10. The `stoplist` file was the default obtained from GTP. Parameter `min_global_freq` was set to 2, so terms appearing only once were excluded, and stemming was enabled. As shown in Table 10, stemming causes a significant – up to 36% – reduction in dictionary size.

For LSI, the matrix was approximated with the leading 100 singular triplets. As described in Sect. 2.2, there are 60 possible combinations for the term weighting and normalization in constructing the term document matrix and 30 possible combinations in constructing the query vector. Taking into account the stemming option, there are 3,600 possible parameter combinations.

Table 10. Dataset statistics for "query answering" experiments

Feature	MEDLINE	CRANFIELD	CISI
Documents	1,033	1,398	1,460
Terms (indexing)	5,735	4,563	5,544
Terms/document	157	189	302
Terms/document (indexing)	71	92	60
tdm nonzeros (%)	0.86	1.27	0.84
# queries	30	225	35
Terms/query	22	19	16
Terms/query (indexing)	11	9	8
Terms (after stemming)	4,241	3,091	3,557
Dictionary size reduction (%)	26	32	36

Table 11. VSM precision

MEDLINE		CRANFIELD		CISI		MEDLINE		CRANFIELD		CISI	
ngc.bp_s	58.45	lgc.nf_s	43.27	lpx.lg_s	24.13	ngc.nf_s	58.03	lxc.ne_s	42.93	apx.ag_s	23.69
lgc.bp_s	58.41	lgc.be_s	43.25	lgx.lp_s	24.13	lfc.bg_s	58.01	lxc.bf_s	42.92	agx.ap_s	23.69
ngc.bf_s	58.36	lgc.ne_s	43.23	lpx.tg_s	23.96	ngc.ne_s	58.00	lxc.nf_s	42.87	apx.ng_s	23.59
npc.bg_s	58.35	lgc.bf_s	43.20	lgx.tp_s	23.96	lgc.be_s	57.97	lgc.np_s	42.87	agx.np_s	23.59
ngc.be_s	58.35	ngc.be_s	43.16	lpx.ag_s	23.93	npc.ng_s	57.81	ngc.le_s	42.82	npx.tg_s	23.44
lgc.bf_s	58.20	ngc.bf_s	43.12	lgx.ap_s	23.93	npx.bg_s	57.81	lxc.lf_s	42.82	ngx.tp_s	23.44
lpc.bg_s	58.17	lxc.be_s	43.03	lpx.ng_s	23.81	ngx.bp_s	57.81	agc.nf_s	42.81	ngx.tf_s	23.41
nec.bg_s	58.17	ngc.nf_s	43.03	lgx.np_s	23.81	agc.bp_s	57.73	lgc.bp_s	42.81	nfx.tg_s	23.41
ngc.np_s	58.15	ngc.ne_s	42.98	apx.lg_s	23.79	lec.bg_s	57.70	ngc.lf_s	42.81	npx.ag_s	23.36
nfc.bg_s	58.15	lgc.le_s	42.96	agx.lp_s	23.79	lgc.ne_s	57.69	lgc.af_s	42.79	ngx.lf_s	23.36
lgc.np_s	58.12	lxc.le_s	42.96	apx.tg_s	23.69	agc.bf_s	57.67	lgc.tf_s	42.79	ngx.lf_s	23.36
lgc.nf_s	58.08	lgc.lf_s	42.95	agx.tp_s	23.69	nec.ng_s	57.67	agc.ne_s	42.76	nfx.lg_s	23.36

Table 12. LSI precision

MEDLINE		CRANFIELD		CISI		MEDLINE		CRANFIELD		CISI	
lfc.bp_s	69.51	aec.bn_s	46.23	aec.lp_s	24.79	lec.np_s	69.07	lpc.bf_s	45.95	lfc.le_s	24.24
lec.bp_s	69.39	lec.bn_s	46.18	aec.np_s	24.66	lfc.nf_s	69.06	lpc.be_s	45.92	lfc.lp_s	24.22
lec.bf_s	69.38	lec.nn_s	46.13	lfc.tf_s	24.45	lpc.nf_s	69.05	lfc.bp_s	45.89	afc.tf_s	24.22
lfc.bf_s	69.33	aec.nn_s	46.09	lfc.lf_s	24.40	lpc.ne_s	69.03	lec.bf_s	45.87	lfc.tp_s	24.19
lpc.bp_s	69.31	lec.ln_s	46.07	lfc.nf_s	24.35	lec.nf_s	69.00	lfc.be_s	45.87	aec.bp_s	24.19
lpc.be_s	69.31	afc.bn_s	46.06	lec.lf_s	24.35	lpx.bp_s	68.99	lfc.nf_s	45.87	lec.bp_s	24.17
lpc.bf_s	69.27	aec.ln_s	46.05	aec.le_s	24.33	aec.bp_s	68.99	aec.an_s	45.82	lec.nf_s	24.16
lfc.be_s	69.27	lec.an_s	46.04	lfc.af_s	24.32	lfx.be_s	68.93	lpc.ne_s	45.82	afc.af_s	24.16
lfc.np_s	69.25	lec.tn_s	46.04	lec.np_s	24.32	lfc.ne_s	68.92	aec.tn_s	45.82	lfc.ne_s	24.16
lpc.np_s	69.16	lfc.bf_s	46.02	lfc.te_s	24.31	aec.bf_s	68.92	afc.an_s	45.79	aec.ne_s	24.14
lec.be_s	69.13	afc.nn_s	45.99	aec.lf_s	24.25	lpx.bf_s	68.92	afc.tn_s	45.79	afc.lf_s	24.14
afc.bp_s	69.09	afc.ln_s	45.96	lfc.ae_s	24.24	afc.bf_s	68.91	lpc.nf_s	45.79	lec.ne_s	24.13

We ran all of them on the aforementioned data collections and recorded the results. It must be noted that this is an exhaustive experiment of considerable magnitude, taking approximately 10h of computation. Tables 11 and 12 list the means of the 25 best precision values obtained amongst all weighting and normalization schemes used for query answering using VSM and LSI. Symbols "_s" and "_ns" indicate the presence or the absence of stemming. Tables 11 and 12 show the performance of LSI for the best weighting and normalization options. First, note that LSI returns good precision, about 19% better than VSM for MEDLINE. The performance of each weighting scheme does not seem to vary across collections. For example, the "logarithmic" local term and the "gfidf" global term weighting schemes appear to return the best precision values for VSM. In the case of LSI, it appears that "logarithmic" local term weighting gives similar results, while "IDF" and "probabilistic inverse" global term weighting return the best performance. Furthermore, precision is generally better with stemming. In view of this and the reduction in dictionary size, stemming appears to be a desirable feature in both VSM and LSI.

Clustering

We next present results concerning the effects of term weighting and stemming on clustering. For our experiments, we used parts of REUT-ALL. We remind the reader that the latter consists of 19,042 documents, 8,654 of which belong to a single topic. We applied TMG in four parts of REUT-ALL, labeled REUTC1, REUTC2, REUTC3, and REUTC4. Each of these consist of documents from 22, 9, 6, and 25 classes, respectively. REUTC1 up to REUTC3 contain an equal number of documents from each class (i.e., 40, 100, and 200, respectively). REUTC4, on the other hand, consists of documents with varying class sizes, ranging from 30 to 300. Table 13 summarizes the features of our datasets. As before, stemming causes again a significant – up to 31% – reduction in dictionary size. As in Sect. 4.2, we tried all possible weighting and normalization options available in TMG and recorded the resulting entropy values for two clustering schemes: PDDP [76], as a representative hierarchical algorithm, based on spectral information, and Spherical k-means (Skmeans) [136] as an interesting partitioning algorithm. Tables 14 and 15 summarize the entropy values using the combinations of the ten weighting and normalization schemes that returned the best results. Skmeans entropy values are about 45% better than PDDP for REUTC2. Stemming and cosine normalization appear to improve the quality of clustering in most cases. Tables 14, and 15 do not identify a specific weighting scheme as best, though "logarithmic" and "alternate log" local and "entropy" and "IDF" global weighting appear to return good results. Moreover, the simple "term frequency" local function appears to return good clustering performance whereas global weighting does not seem to improve matters.

5 Conclusions

We have outlined the design and implementation of TMG, a novel MATLAB toolbox for the construction of tdms from text collections presented in the

Table 13. Document collections used in clustering experiments

Feature	REUT-ALL	REUTC1	REUTC2	REUTC3	REUTC4
Documents	19,042	880	900	1,200	2,936
Terms (indexing)	21,698	4,522	4,734	5,279	8,846
Terms/document	145	179	180	175	180
Terms/document (indexing)	69	81	83	81	85
tdm nonzeros (%)	0.22	0.20	0.20	1.06	0.66
Terms (after stemming)	15,295	3,228	3,393	3,691	6,068
Dictionary size reduction (%)	30	29	28	30	31

Table 14. Entropy values for PDDP

REUT1		REUT2		REUT3		REUT4	
tpc_s	1.46	lec_ns	1.11	aec_s	0.85	afc_ns	1.63
tec_s	1.54	afc_ns	1.13	lec_s	0.90	tfc_s	1.64
tfc_s	1.58	aec_ns	1.15	tec_s	0.92	aec_ns	1.67
tec_ns	1.59	lec_s	1.17	tec_ns	0.93	afc_s	1.68
lec_s	1.61	lfc_ns	1.18	bxc_ns	0.96	lfc_ns	1.68
aec_s	1.61	lfc_s	1.19	tfc_s	0.96	tec_ns	1.69
tpc_ns	1.63	aec_s	1.20	lec_ns	0.97	tec_s	1.69
aec_ns	1.66	afc_s	1.24	afc_s	0.98	aec_s	1.72
afc_s	1.67	tfc_ns	1.26	aec_ns	1.01	tpc_ns	1.72
afc_ns	1.67	lgc_ns	1.29	afc_ns	1.01	lec_s	1.73

Table 15. Entropy values for Skmeans

REUT1		REUT2		REUT3		REUT4	
tpc_s	1.18	axc_s	0.61	bxc_ns	0.66	lec_s	0.96
tpc_ns	1.23	aec_s	0.73	lec_ns	0.67	tfc_s	0.98
tfc_s	1.28	lec_ns	0.73	lxc_ns	0.73	tec_s	0.99
tec_s	1.30	lxc_s	0.73	axc_s	0.74	afc_ns	1.03
afc_s	1.31	tfc_s	0.73	bxc_s	0.74	aec_s	1.03
tec_ns	1.31	nxc_s	0.74	bgc_s	0.74	lec_ns	1.04
lec_ns	1.33	lxc_ns	0.75	tec_ns	0.75	afc_s	1.04
axc_s	1.35	axc_ns	0.76	nxc_ns	0.78	tec_ns	1.06
afc_ns	1.35	tec_ns	0.76	bgc_ns	0.78	apc_s	1.06
ngc_s	1.36	bgc_s	0.76	tpc_ns	0.79	tfc_ns	1.07

form of ASCII text files and directories. Our motivation was to facilitate users, such as researchers and educators in computational linear algebra, who use MATLAB to build algorithms for textual information retrieval and are interested in the rapid preparation of test data. By using TMG one avoids the extra steps necessary to convert or interface with data produced by other systems. TMG returns (albeit slower) results comparable with those produced by GTP, a popular C++ package for IR using LSI. TMG also allows one to conduct stemming by means of a well-known variation of Porter's algorithm and provides facilities for the maintenance and incremental construction of term-document collections. We presented examples of use of TMG in various settings and data collections, including BIBBENCH, a new dataset consisting of data in BIBTEX format. The flexibility of TMG allowed us extensive experimentation with various combinations of term weighting and normalization schemes and stemming. The tool is publicly available via a simple request. We are currently working in enabling the tool to process a variety of other document types as well as in distributed implementations. We intend to exploit the facilities for integer and single-precision arithmetic of MATLAB 7.0 as well as compression techniques to produce a more efficient implementation.

Acknowledgments

We thank Jacob Kogan and Charles Nicholas for inviting us to contribute to this volume. TMG was conceived after a motivating discussion with Andrew Knyazev regarding a collection of MATLAB tools we had put together to aid in our clustering experiments. We thank Michael Berry for discussions and for including the software in the LSI web site [5], Efi Kokiopoulou and Constantine Bekas for many helpful suggestions, Dan Boley for his help regarding preprocessing in PDDP. We thank Inderjit Dhillon, Pavel Berkhin and the editors for letting us access an early version of this volume's BIBTEX source to use in our experiments and Michael Velgakis for assistance regarding the BEC dataset. We thank Elias Houstis for his help in the initial phases of this research and for providing us access to MATLAB 7.0. Special thanks are due to many of the users for their constructive comments regarding TMG. This research was supported in part by a University of Patras "Karatheodori" grant. The first author was also supported by a Bodossaki Foundation graduate fellowship.

Appendix

Availability

TMG and its documentation are available from the following URL:

> http://scgroup.hpclab.ceid.upatras.gr/scgroup/Projects/TMG/

Assuming that the file has been downloaded and saved into a directory that is already in the MATLAB path or made to be that way by executing the MATLAB command (addpath('directory')), TMG is ready for use. To process Adobe PDF and POSTSCRIPT files, the Ghostscript utility ps2ascii and Ghostscript's compiler must be made available in the path. TMG checks the availability of this utility and uses it if available, otherwise processes the next document. It is also recommended, before TMG, to use ps_pdf2ascii, that is our MATLAB interface to ps2ascii. This checks the resulting ASCII file as the results are not always desired. The user can also easily edit ps_pdf2ascii to insert additional filters so that the system can process additional file formats (e.g., detex for TEX files). MATLAB version 6.5 or higher is assumed.

Criterion Functions for Clustering on High-Dimensional Data

Y. Zhao and G. Karypis

Summary. In recent years, we have witnessed a tremendous growth in the volume of text documents available on the Internet, digital libraries, news sources, and company-wide intranets. This has led to an increased interest in developing methods that can help users to effectively navigate, summarize, and organize this information with the ultimate goal of helping them to find what they are looking for. Fast and high-quality document clustering algorithms play an important role toward this goal as they have been shown to provide both an intuitive navigation/browsing mechanism by organizing large amounts of information into a small number of meaningful clusters as well as to greatly improve the retrieval performance either via cluster-driven dimensionality reduction, term-weighting, or query expansion. This ever-increasing importance of document clustering and the expanded range of its applications led to the development of a number of new and novel algorithms with different complexity-quality trade-offs. Among them, a class of clustering algorithms that have relatively low computational requirements are those that treat the clustering problem as an optimization process, which seeks to maximize or minimize a particular *clustering criterion function* defined over the entire clustering solution.

This chapter provides empirical and theoretical comparisons of the performance of a number of widely used criterion functions in the context of partitional clustering algorithms for high-dimensional datasets. The comparisons consist of a comprehensive experimental evaluation involving 15 different datasets, as well as an analysis of the characteristics of the various criterion func-break tions and their effect on the clusters they produce. Our experimental results show that there is a set of criterion functions that consistently outperform the rest, and that some of the newly proposed criterion functions lead to the best overall results. Our theoretical analysis of the criterion function shows that their relative performance of the criterion functions depends on: (i) the degree to which they can correctly operate when the clusters are of different tightness, and (ii) the degree to which they can lead to reasonably balanced clusters.

1 Introduction

The topic of clustering has been extensively studied in many scientific disciplines and a variety of different algorithms have been developed [61, 76, 105, 126, 139, 207, 208, 241, 260, 271, 323, 349, 404, 409, 450]. Two recent surveys on the topics [218, 245] offer a comprehensive summary of the different applications and algorithms. These algorithms can be categorized along different dimensions based either on the underlying methodology of the algorithm, leading to *agglomerative* or *partitional* approaches, or on the structure of the final solution, leading to *hierarchical* or *nonhierarchical* solutions.

In recent years, various researchers have recognized that partitional clustering algorithms are well suited for clustering large document datasets due to their relatively low computational requirements [121, 303, 407]. A key characteristic of many partitional clustering algorithms is that they use a global criterion function whose optimization drives the entire clustering process. For some of these algorithms the criterion function is implicit (e.g., Principal Direction Divisive Partitioning (PDDP) [76]), whereas for other algorithms (e.g., k-means [323], Cobweb [164], and Autoclass [105]) the criterion function is explicit and can be easily stated. This latter class of algorithms can be thought of as consisting two key components. The first is the criterion function that the clustering solution optimizes, and the second is the actual algorithm that achieves this optimization.

The focus of this chapter is to study the suitability of different criterion functions to the problem of clustering document datasets. In particular, we evaluate a total of seven criterion functions that measure various aspects of intracluster similarity, intercluster dissimilarity, and their combinations. These criterion functions utilize different views of the underlying collection by either modeling the documents as vectors in a high-dimensional space or by modeling the collection as a graph. We experimentally evaluated the performance of these criterion functions using 15 different datasets obtained from various sources. Our experiments show that different criterion functions do lead to substantially different results and that there is a set of criterion functions that produce the best clustering solutions.

Our analysis of the different criterion functions shows that their overall performance depends on the degree to which they can correctly operate when the dataset contains clusters of different tightness (i.e., they contain documents whose average pairwise similarities are different) and the degree to which they can produce balanced clusters. Moreover, our analysis also shows that the sensitivity to the difference in the cluster tightness can also explain an outcome of our study (which was also observed in earlier results reported in [407]), that for some clustering algorithms the solution obtained by performing a sequence of repeated bisections (RB) is better (and for some criterion functions by a considerable amount) than the solution obtained by computing the clustering directly. When the solution is computed via RB, the tightness difference between the two clusters that are discovered is in general smaller than that between all the clusters. As a result, criterion functions that cannot handle well variation in cluster tightness tend to perform substantially better when used to compute the clustering via RB.

The rest of this chapter is organized as follows. Section 2 provides some information on the document representation and similarity measure used in our study. Section 3 describes the different criterion functions and the algorithms used to optimize them. Section 4 provides the detailed experimental evaluation of the various criterion functions. Section 5 analyzes the different criterion functions and explains their performance. Finally, Sect. 6 provides some concluding remarks.

2 Preliminaries

2.1 Document Representation

The various clustering algorithms described in this chapter represent each document using the well-known *term frequency–inverse document frequency* (tf–idf) vector-space model [377]. In this model, each document d is considered to be a vector in the term-space and is represented by the vector

$$d_{\text{tfidf}} = (tf_1 \log(n/df_1), tf_2 \log(n/df_2), \ldots, tf_m \log(n/df_m)),$$

where tf_i is the frequency of the ith term (i.e., term frequency), n is the total number of documents, and df_i is the number of documents that contain the ith term (i.e., document frequency). To account for documents of different lengths, the length of each document vector is normalized so that it is of unit length. In the rest of the chapter, we will assume that the vector representation for each document has been weighted using tf–idf and normalized so that it is of unit length.

2.2 Similarity Measures

Two important ways have been proposed to compute the similarity between two documents d_i and d_j. The first method is based on the commonly used [377] cosine function

$$\cos(d_i, d_j) = d_i{}^t d_j / (\|d_i\| \|d_j\|),$$

and since the document vectors are of unit length, it simplifies to $d_i{}^t d_j$. The second method computes the similarity between the documents using the Euclidean distance $\text{dis}(d_i, d_j) = \|d_i - d_j\|$. Note that besides the fact that one measures similarity and the other measures distance, these measures are quite similar to each other because the document vectors are of unit length.

2.3 Definitions

Throughout this chapter we use the symbols n, m, and k to denote the number of documents, the number of terms, and the number of clusters, respectively. We use the symbol S to denote the set of n documents to be clustered, $S_1, S_2, \ldots, S_k$ to denote each one of the k clusters, and $n_1, n_2, \ldots, n_k$ to denote their respective sizes. Given a set A of documents and their corresponding vector representations, we define the *composite* vector D_A to be $D_A = \sum_{d \in A} d$ and the *centroid* vector C_A to be $C_A = D_A/|A|$.

The composite vector D_A is nothing more than the sum of all documents vectors in A, and the centroid C_A is nothing more than the vector obtained by averaging the weights of the various terms present in the documents of A. Note that even though the document vectors are of length one, the centroid vectors will not necessarily be of unit length.

2.4 Vector Properties

By using the cosine function as the measure of similarity between documents, we can take advantage of a number of properties involving the composite and centroid vectors of a set of documents. In particular, if S_i and S_j are two sets of unit-length documents containing n_i and n_j documents, respectively, and D_i, D_j and C_i, C_j are their corresponding composite and centroid vectors then the following is true:

1. The sum of the pairwise similarities between the documents in S_i and the document in S_j is equal to $D_i{}^t D_j$. That is,

$$\sum_{d_q \in D_i, d_r \in D_j} \cos(d_q, d_r) = \sum_{d_q \in D_i, d_r \in D_j} d_q{}^t d_r = D_i{}^t D_j. \tag{1}$$

2. The sum of the pairwise similarities between the documents in S_i is equal to $\|D_i\|^2$. That is,

$$\sum_{d_q, d_r \in D_i} \cos(d_q, d_r) = \sum_{d_q, d_r \in D_i} d_q{}^t d_r = D_i{}^t D_i = \|D_i\|^2. \tag{2}$$

Note that this equation includes the pairwise similarities involving the same pairs of vectors.

3 Document Clustering

At a high level the problem of clustering is defined as follows. Given a set S of n documents, we would like to partition them into a predetermined number of k subsets $S_1, S_2, \ldots, S_k$, such that the documents assigned to each subset are more similar to each other than those assigned to different subsets.

Table 1. Clustering criterion functions

$$\mathcal{I}_1 \quad \text{maximize} \quad \sum_{r=1}^{k} n_r \left(\frac{1}{n_r^2} \sum_{d_i, d_j \in S_r} \cos(d_i, d_j) \right) = \sum_{r=1}^{k} \frac{\|D_r\|^2}{n_r} \qquad (3)$$

$$\mathcal{I}_2 \quad \text{maximize} \quad \sum_{r=1}^{k} \sum_{d_i \in S_r} \cos(d_i, C_r) = \sum_{r=1}^{k} \|D_r\| \qquad (4)$$

$$\mathcal{E}_1 \quad \text{minimize} \quad \sum_{r=1}^{k} n_r \cos(C_r, C) \Leftrightarrow \text{minimize} \quad \sum_{r=1}^{k} n_r \frac{D_r{}^t D}{\|D_r\|} \qquad (5)$$

$$\mathcal{H}_1 \quad \text{maximize} \quad \frac{\mathcal{I}_1}{\mathcal{E}_1} \Leftrightarrow \text{minimize} \quad \frac{\sum_{r=1}^{k} \|D_r\|^2 / n_r}{\sum_{r=1}^{k} n_r D_r{}^t D / \|D_r\|} \qquad (6)$$

$$\mathcal{H}_2 \quad \text{maximize} \quad \frac{\mathcal{I}_2}{\mathcal{E}_1} \Leftrightarrow \text{minimize} \quad \frac{\sum_{r=1}^{k} \|D_r\|}{\sum_{r=1}^{k} n_r D_r{}^t D / \|D_r\|} \qquad (7)$$

$$\mathcal{G}_1 \quad \text{minimize} \quad \sum_{r=1}^{k} \frac{\text{cut}(S_r, S - S_r)}{\sum_{d_i, d_j \in S_r} \cos(d_i, d_j)} = \sum_{r=1}^{k} \frac{D_r{}^t (D - D_r)}{\|D_r\|^2} \qquad (8)$$

$$\mathcal{G}_2 \quad \text{minimize} \quad \sum_{r=1}^{k} \frac{\text{cut}(V_r, V - V_r)}{W(V_r)} \qquad (9)$$

As discussed in the introduction, our focus is to study the suitability of various clustering criterion functions in the context of partitional document clustering algorithms. Consequently, given a particular clustering criterion function $\mathcal{C}$, the clustering problem is to compute a k-way clustering solution such that the value of $\mathcal{C}$ is optimized. In the rest of this section, we first present a number of different criterion functions that can be used to both evaluate and drive the clustering process, followed by a description of the algorithms that were used to perform their optimization.

3.1 Clustering Criterion Functions

Our study involves a total of seven different clustering criterion functions that are summarized in Table 1. These functions optimize various aspects of intracluster similarity, intercluster dissimilarity, and their combinations, and represent some of the most widely used criterion functions for document clustering.

Internal Criterion Functions

This class of clustering criterion functions focuses on producing a clustering solution that optimizes a particular criterion function that is defined over the documents that are part of each cluster and does not take into account the documents assigned to different clusters. Due to this intracluster view of the clustering process, we refer to these criterion functions as *internal*.

The first internal criterion function that we study maximizes the sum of the average pairwise similarities between the documents assigned to each cluster weighted according to the size of each cluster, and has been used successfully for clustering document datasets [363]. Specifically, if we use the cosine function to measure the similarity between documents, then we wish the clustering solution to optimize the following criterion function:

$$\text{maximize } \mathcal{I}_1 = \sum_{r=1}^{k} n_r \left(\frac{1}{n_r^2} \sum_{d_i, d_j \in S_r} \cos(d_i, d_j) \right). \tag{10}$$

By using (2), the above formula can be rewritten as:

$$\mathcal{I}_1 = \sum_{r=1}^{k} \frac{\|D_r\|^2}{n_r}.$$

Note that our definition of $\mathcal{I}_1$ includes the self-similarities between the documents of each cluster. The $\mathcal{I}_1$ criterion function is similar to that used in the context of hierarchical agglomerative clustering that uses the group-average heuristic to determine the pair of clusters to merge next.

The second criterion function that we study is used by the popular vector-space variant of the K-means algorithm [121, 136, 303, 407]. In this algorithm each cluster is represented by its centroid vector and the goal is to find the clustering solution that maximizes the similarity between each document and the centroid of the cluster that is assigned to it. Specifically, if we use the cosine function to measure the similarity between a document and a centroid, then the criterion function becomes the following:

$$\text{maximize } \mathcal{I}_2 = \sum_{r=1}^{k} \sum_{d_i \in S_r} \cos(d_i, C_r). \tag{11}$$

This formula can be rewritten as follows:

$$\mathcal{I}_2 = \sum_{r=1}^{k} \sum_{d_i \in S_r} \frac{d_i{}^t C_r}{\|C_r\|} = \sum_{r=1}^{k} \frac{D_r{}^t C_r}{\|C_r\|} = \sum_{r=1}^{k} \frac{D_r{}^t D_r}{\|D_r\|} = \sum_{r=1}^{k} \|D_r\|.$$

Comparing the $\mathcal{I}_2$ criterion function with $\mathcal{I}_1$ we can see that the essential difference between these criterion functions is that $\mathcal{I}_2$ scales the within-cluster similarity by the $\|D_r\|$ term as opposed to n_r term used by $\mathcal{I}_1$. The term $\|D_r\|$ is nothing more than the square root of the pairwise similarity between all the documents in S_r, and will tend to emphasize the importance of clusters (beyond the $\|D_r\|^2$ term) whose documents have smaller pairwise similarities compared to clusters with higher pairwise similarities. Also note that if the similarity between a document and the centroid vector of its cluster is defined as just the dot-product of these vectors, then we will get back the $\mathcal{I}_1$ criterion function.

Finally, the last internal criterion function that we study is that used by the traditional K-means algorithm. This criterion function uses the Euclidean distance to determine the documents that should be clustered together, and determines the overall quality of the clustering solution by using the *sum-of-squared-errors* function. In particular, this criterion is defined as follows:

$$\text{minimize } \mathcal{I}_3 = \sum_{r=1}^{k} \sum_{d_i \in S_r} \| d_i - C_r \|^2. \tag{12}$$

Note that by some simple algebraic manipulations [144], the above equation can be rewritten as:

$$\mathcal{I}_3 = \sum_{r=1}^{k} \frac{1}{n_r} \sum_{d_i, d_j \in S_r} \| d_i - d_j \|^2, \tag{13}$$

which is similar in nature to the $\mathcal{I}_1$ criterion function but instead of using similarities it is expressed in terms of squared distances. Since the cosine and Euclidean distance functions are similar to each other, $\mathcal{I}_3$ exhibits similar characteristics with $\mathcal{I}_1$. To see this, from (13), using some basic trigonometric manipulations we have that

$$\| d_i - d_j \|^2 = \sin^2(d_i, d_j) + (1 - \cos(d_i, d_j))^2 = 2(1 - \cos(d_i, d_j)).$$

Using this relation, (13) can be rewritten as:

$$\mathcal{I}_3 = \sum_{r=1}^{k} \frac{1}{n_r} \sum_{d_i, d_j \in S_r} 2(1 - \cos(d_i, d_j))$$

$$= 2 \left(\sum_{r=1}^{k} n_r - \sum_{r=1}^{k} \frac{1}{n_r} \sum_{d_i, d_j \in S_r} \cos(d_i, d_j) \right)$$

$$= 2(n - \mathcal{I}_1).$$

Thus, minimizing $\mathcal{I}_3$ is the same as maximizing $\mathcal{I}_1$. Therefore, we will not discuss $\mathcal{I}_3$ any further.

External Criterion Functions

Unlike internal criterion functions, external criterion functions derive the clustering solution by focusing on optimizing a function that is based on how the various clusters are different from each other. Due to this intercluster view of the clustering process we refer to these criterion functions as *external*.

It is quite hard to define external criterion functions that lead to meaningful clustering solutions. For example, it may appear that an intuitive external function may be derived by requiring that the centroid vectors of the different clusters are as mutually orthogonal as possible, i.e., they contain documents that share very few terms across the different clusters. However, for many problems this criterion function has trivial solutions that can be achieved by assigning to the first $k-1$ clusters a single document that shares very few terms with the rest, and then assigning the rest of the documents to the kth cluster.

For this reason, the external function that we will study tries to separate the documents of each cluster from the entire collection, as opposed trying to separate the documents among the different clusters. In particular, our external criterion function is defined as

$$\text{minimize } \sum_{r=1}^{k} n_r \cos(C_r, C), \tag{14}$$

where C is the centroid vector of the entire collection. From this equation, we can see that we try to minimize the cosine between the centroid vector of each cluster to the centroid vector of the entire collection. By minimizing the cosine we essentially try to increase the angle between them as much as possible. Also note that the contribution of each cluster is weighted based on the cluster size, so that larger clusters will weight heavier in the overall clustering solution. This external criterion function was motivated by multiple discriminant analysis and is similar to minimizing the trace of the between-cluster scatter matrix [144].

(14) can be rewritten as

$$\sum_{r=1}^{k} n_r \cos(C_r, C) = \sum_{r=1}^{k} n_r \frac{C_r{}^t C}{\|C_r\|\|C\|} = \sum_{r=1}^{k} n_r \frac{D_r{}^t D}{\|D_r\|\|D\|}$$
$$= \frac{1}{\|D\|} \left(\sum_{r=1}^{k} n_r \frac{D_r{}^t D}{\|D_r\|} \right),$$

where D is the composite vector of the entire document collection. Note that since $1/\|D\|$ is constant irrespective of the clustering solution the criterion function can be restated as:

$$\text{minimize } \mathcal{E}_1 = \sum_{r=1}^{k} n_r \frac{D_r{}^t D}{\|D_r\|}. \tag{15}$$

As we can see from (15), even though our initial motivation was to define an external criterion function, because we used the cosine function to measure the separation between the cluster and the entire collection, the criterion function does take into account the within-cluster similarity of the documents (due to the $\|D_r\|$ term). Thus, $\mathcal{E}_1$ is actually a hybrid criterion function that combines both external as well as internal characteristics of the clusters.

Another external criterion function can be defined with respect to the Euclidean distance function and the squared errors of the centroid vectors as follows:

$$\text{maximize } \mathcal{E}_2 = \sum_{r=1}^{k} n_r \|C_r - C\|^2. \tag{16}$$

However, it can be shown that maximizing $\mathcal{E}_\in$ is identical to minimizing $\mathcal{I}_3$ [144], and we will not consider it any further.

Hybrid Criterion Functions

The various criterion functions we described so far focus only on optimizing a single criterion function that is either defined in terms of how documents assigned to each cluster are related together, or how the documents assigned to each cluster are related with the entire collection. In the first case, they try to maximize various measures of similarity over the documents in each cluster, and in the second case, they try to minimize the similarity between the cluster's documents and the collection. However, the various clustering criterion functions can be combined to define a set of *hybrid* criterion functions that simultaneously optimize multiple individual criterion functions.

In our study, we focus on two hybrid criterion function that are obtained by combining criterion $\mathcal{I}_1$ with $\mathcal{E}_1$, and $\mathcal{I}_2$ with $\mathcal{E}_1$, respectively. Formally, the first criterion function is

$$\text{maximize } \mathcal{H}_1 = \frac{\mathcal{I}_1}{\mathcal{E}_1} = \frac{\sum_{r=1}^{k} \|D_r\|^2 / n_r}{\sum_{r=1}^{k} n_r D_r{}^t D / \, nrm D_r}, \tag{17}$$

and the second is

$$\text{maximize } \mathcal{H}_2 = \frac{\mathcal{I}_2}{\mathcal{E}_1} = \frac{\sum_{r=1}^{k} \|D_r\|}{\sum_{r=1}^{k} n_r D_r{}^t D / \|D_r\|}. \tag{18}$$

Note that since $\mathcal{E}_1$ is minimized, both $\mathcal{H}_1$ and $\mathcal{H}_2$ need to be maximized as they are inversely related to $\mathcal{E}_1$.

Graph-Based Criterion Functions

The various criterion functions that we described so far, view each document as a multidimensional vector. An alternate way of viewing the relations between the documents is to use graphs. In particular, two types of graphs have been proposed for modeling the document in the context of clustering. The first graph is nothing more than the graph obtained by computing the pairwise similarities between the documents, and the second graph is obtained by viewing the documents and the terms as a bipartite graph [52, 128, 455]. By viewing the documents in this fashion, a number of edge-cut-based criterion functions (i.e., *graph-based* criterion functions) can be used to cluster document datasets [109, 128, 139, 212, 395, 455]. $\mathcal{G}_1$ and $\mathcal{G}_2$ ((19) and (20)) are two such criterion functions that are defined on the similarity and bipartite graphs, respectively.

Given a collection of n documents S, the similarity graph G_s is obtained by modeling each document as a vertex, and having an edge between each pair of vertices whose weight is equal to the similarity between the corresponding documents. By viewing the documents in this fashion, a number of internal, external, or combined criterion functions can be defined that measure the overall clustering quality. In our study we investigate one such criterion function called MinMaxCut, that was proposed recently [139]. MinMaxCut falls under the category of criterion functions that combine both the internal and me external views of the clustering process and is defined as [139]

$$\text{minimize} \quad \sum_{r=1}^{k} \frac{\text{cut}(S_r, S - S_r)}{\sum_{d_i, d_j \in S_r} \text{sim}(d_i, d_j)},$$

where $\text{cut}(S_r, S - S_r)$ is the edge cut between the vertices in S_r to the rest of the vertices in the graph $S - S_r$. The edge cut between two sets of vertices A and B is defined to be the sum of the edges connecting vertices in A to vertices in B. The motivation behind this criterion function is that the clustering process can be viewed as that of partitioning the documents into groups by minimizing the edge cut of each partition. However, for reasons similar to those discussed in Sect. 3.1, such an external criterion may have trivial solutions, and for this reason each edge cut is scaled by the sum of the internal edges. As shown in [139], this scaling leads to better balanced clustering solutions. If we use the cosine function to measure the similarity between the documents, and (1) and (2), then the above criterion function can be rewritten as

$$\sum_{r=1}^{k} \frac{\sum_{d_i \in S_r, d_j \in S - S_r} \cos(d_i, d_j)}{\sum_{d_i, d_j \in S_r} \cos(d_i, d_j)} = \sum_{r=1}^{k} \frac{D_r{}^t (D - D_r)}{\|D_r\|^2} = \left(\sum_{r=1}^{k} \frac{D_r{}^t D}{\|D_r\|^2} \right) - k,$$

and since k is constant, the criterion function can be simplified to

$$\text{minimize} \ \mathcal{G}_1 = \sum_{r=1}^{k} \frac{D_r{}^t D}{\|D_r\|^2}. \tag{19}$$

An alternate graph model views the various documents and their terms as a bipartite graph $G_b = (V, E)$, where V consists of two sets V_d and V_t. The vertex set V_d corresponds to the documents whereas the vertex set V_t corresponds to the terms. In this model, if the ith document contains the jth term, there is an edge connecting the corresponding ith vertex of V_d to the jth vertex of V_t. The weights of these edges are set using the tf–idf model discussed in Sect. 2. Given such a bipartite graph, the problem of clustering can be viewed as that of computing a simultaneous partitioning of the documents and the terms so that a criterion function defined on the edge cut is optimized. In our study, we focus on a particular edge cut-based criterion function called the normalized cut, which had been recently used in the context of this bipartite graph model for document clustering [128, 455]. The normalized cut criterion function is defined as

$$\text{minimize } \mathcal{G}_2 = \sum_{r=1}^{k} \frac{\text{cut}(V_r, V - V_r)}{W(V_r)}, \tag{20}$$

where V_r is the set of vertices assigned to the rth cluster, and $W(V_r)$ is the sum of the weights of the adjacency lists of the vertices assigned to the rth cluster. Note that the rth cluster will contain vertices from both the V_d and V_t, i.e., both documents as well as terms. The key motivation behind this representation and criterion function is to compute a clustering that groups together documents as well as the terms associated with these documents. Also, note that the various $W(V_r)$ quantities are used primarily as normalization factors, to ensure that the optimization of the criterion function does not lead to trivial solutions. Its purpose is similar to the $\|D_r\|^2$ factor used in $\mathcal{G}_1$ (19).

3.2 Criterion Function Optimization

There are many techniques that can be used to optimize the criterion functions described in Sect. 3.1 They include relatively simple greedy schemes, iterative schemes with varying degree of hill-climbing capabilities, and powerful but computationally expensive spectral-based optimizers [76,105,128,165,261, 323,333,454,455]. Despite this wide range of choices, in our study, the various criterion functions were optimized using a simple and obvious greedy strategy. This was primarily motivated by our experience with document datasets (and similar results presented in [385]), which showed that greedy-based schemes (when run multiple times) produce comparable results to those produced by more sophisticated optimization algorithms for the range of the number of clusters that we used in our experiments. Nevertheless, the choice of the optimization methodology can potentially impact the relative performance of the various criterion functions, since that performance may depend on the optimizer [165]. However, as we will see later in Sect. 5, our analysis of the criterion functions correlates well with our experimental results, suggesting that the choice of the optimizer does not appear to be biasing the experimental comparisons.

Our greedy optimizer computes the clustering solution by first obtaining an initial k-way clustering and then applying an iterative refinement algorithm to further improve it. During initial clustering, k documents are randomly selected to form the *seeds* of the clusters and each document is assigned to the cluster corresponding to its most similar seed. This approach leads to an initial clustering solution for all but the $\mathcal{G}_2$ criterion function as it does not produce an initial partitioning for the vertices corresponding to the terms (V_t). The initial partitioning of V_t is obtained by assigning each term v to the partition that is most connected with. The iterative refinement strategy that we used is based on the *incremental* refinement scheme described in [144]. During each iteration, the documents are visited in a random order and each document is moved to the cluster that leads to the highest improvement in the value of the criterion function. If no such cluster exists, then the document does not move. The refinement phase ends, as soon as an iteration is performed in which no documents were moved between clusters. Note that in the case of $\mathcal{G}_2$, the refinement algorithm alternates between document-vertices and term-vertices [288].

The algorithms used during the refinement phase are greedy in nature, they are not guaranteed to converge to a global optimum, and the local optimum solution they obtain depends on the particular set of seed documents that were selected to obtain the initial clustering. To eliminate some of this sensitivity, the overall process is repeated a number of times. That is, we compute N different clustering solutions (i.e., initial clustering followed by cluster refinement), and the one that achieves the best value for the particular criterion function is kept. In all our experiments, we used $N = 10$. For the rest of this discussion when we refer to a clustering solution, we will mean the solution that was obtained by selecting the best (with respect to the value of the respective criterion function) out of these N potentially different solutions.

4 Experimental Results

We experimentally evaluated the performance of the different clustering criterion functions on a number of different datasets. In the rest of this section we first describe the various datasets and our experimental methodology, followed by a description of the experimental results.

4.1 Document Collections

In our experiments, we used a total of 15 datasets [1] whose general characteristics and sources are summarized in Table 2. The smallest of these datasets contained 878 documents and the largest contained 11,162 documents. To ensure diversity in the datasets, we obtained them from different sources. For all datasets we used a stop list to remove common words and the words were stemmed using Porter's suffix-stripping algorithm [362]. Moreover, any term that occurs in fewer than two documents was eliminated.

4.2 Experimental Methodology and Metrics

For each one of the different datasets we obtained a 5-way, 10-way, 15-way, and 20-way clustering solution that optimized the various clustering criterion

Table 2. Summary of datasets used to evaluate the various clustering criterion functions

Data	Source	# of documents	# of terms	# of classes
classic	CACM/CISI/CRANFIELD/MEDLINE [2]	7,089	12,009	4
fbis	FBIS (TREC-5 [423])	2,463	12,674	17
hitech	San Jose Mercury (TREC, TIPSTER Vol. 3)	2,301	13,170	6
reviews	San Jose Mercury (TREC, TIPSTER Vol. 3)	4,069	23,220	5
sports	San Jose Mercury (TREC, TIPSTER Vol. 3)	8,580	18,324	7
la12	LA Times (TREC-5 [423])	6,279	21,604	6
new3	TREC-5 & TREC-6 [423]	9,558	36,306	44
tr31	TREC-5 & TREC-6 [423]	927	10,128	7
tr41	TREC-5 & TREC-6 [423]	878	7,454	10
ohscal	OHSUMED-233445 [228]	11,162	11,465	10
re0	Reuters-21578 [313]	1,504	2,886	13
re1	Reuters-21578 [313]	1,657	3,758	25
k1a	WebACE [216]	2,340	13,879	20
k1b	WebACE [216]	2,340	13,879	6
wap	WebACE [216]	1,560	8,460	20

functions shown in Table 1. The quality of a clustering solution was evaluated using the **entropy** measure, which is based on how the various classes of documents are distributed within each cluster. Given a particular cluster S_r of size n_r, the entropy of this cluster is defined to be

$$E(S_r) = -\frac{1}{\log q} \sum_{i=1}^{q} \frac{n_r^i}{n_r} \log \frac{n_r^i}{n_r},$$

where q is the number of classes in the dataset and n_r^i is the number of documents of the ith class that were assigned to the rth cluster. The entropy of the entire solution is defined to be the sum of the individual cluster entropies weighted according to the cluster size, i.e.,

$$\text{Entropy} = \sum_{r=1}^{k} \frac{n_r}{n} E(S_r).$$

A perfect clustering solution will be the one that leads to clusters that contain documents from only a single class, in which case the entropy will be 0. In general, the smaller the entropy values, the better the clustering solution is.

To eliminate any instances where a particular clustering solution for a particular criterion function gets trapped into a bad local optimum, in all our experiments we found ten different clustering solutions. As discussed in Sect. 3.2 each of these ten clustering solutions correspond to the best solution (in terms of the respective criterion function) out of ten different initial partitioning and refinement phases. As a result, for each particular value of k and criterion function we generated 100 different clustering solutions. The overall number of experiments that we performed was $3 \times 100 \times 4 \times 8 \times 15$ = 144,000, which were completed in about 8 days on a Pentium III@600MHz workstation.

One of the problems associated with such large-scale experimental evaluation is that of summarizing the results in a meaningful and unbiased fashion.

Our summarization is done as follows. For each dataset and value of k, we divided the entropy obtained by a particular criterion function by the smallest entropy obtained for that particular dataset and value of k over the different criterion functions. These ratios represent the degree to which a particular criterion function performed worse than the best criterion function for that dataset and value of k. These ratios are less sensitive to the actual entropy values and the particular value of k. We refer to these ratios as *relative entropies*. Now, for each criterion function and value of k we averaged these relative entropies over the various datasets. A criterion function that has an *average relative entropy* close to 1.0 indicates that this function did the best for most of the datasets. On the other hand, if the average relative entropy is high, then this criterion function performed poorly. In addition to these numerical averages, we evaluated the statistical significance of the relative performance of the criterion functions using a paired-t test [127] based on the original entropies for each dataset. The original entropy values for all the experiments presented in this chapter can be found in [462].

4.3 Evaluation of Direct k-Way Clustering

Our first set of experiments was focused on evaluating the quality of the clustering solutions produced by the various criterion functions when they were used to compute a k-way clustering solution directly. The values for the average relative entropies for the 5-way, 10-way, 15-way, and 20-way clustering solutions are shown in Table 3. The row labeled "Avg" contains the average of these averages over the four sets of solutions. Furthermore, the last column shows the relative ordering of the different schemes using the paired-t test.

From these results we can see that the $\mathcal{I}_1$ and the $\mathcal{G}_2$ criterion functions lead to clustering solutions that are consistently worse (in the range of 19–35%) than the solutions obtained using the other criterion functions. On the other hand, the $\mathcal{I}_2$, $\mathcal{H}_2$, and $\mathcal{H}_1$ criterion functions lead to the best solutions irrespective of the number of clusters. Over the entire set of experiments, these methods are either the best or always within 2% of the best solution.

Table 3. Average relative entropies for the clustering solutions obtained via direct k-way clustering and their statistical significance. Underlined entries represent the best performing scheme in terms of average relative entropies. Note that "$\ll$" indicates that schemes on the right are significantly better than the schemes on the left, and "( )" indicates that the relationship is not significant. The order of the schemes within parentheses represent the order of the weak relationship

k	$\mathcal{I}_1$	$\mathcal{I}_2$	$\mathcal{E}_1$	$\mathcal{H}_1$	$\mathcal{H}_2$	$\mathcal{G}_1$	$\mathcal{G}_2$	Statistical significance test, p-value=0.05
5	1.361	1.041	1.044	1.069	**1.033**	1.092	1.333	$(\mathcal{I}_1, \mathcal{G}_2) \ll (\mathcal{G}_1, \mathcal{H}_1) \ll (\mathcal{E}_1, \mathcal{I}_2, \mathcal{H}_2)$
10	1.312	1.042	1.069	**1.035**	1.040	1.148	1.380	$\mathcal{G}_2 \ll \mathcal{I}_1 \ll \mathcal{G}_1 \ll (\mathcal{E}_1, \mathcal{H}_2, \mathcal{H}_1, \mathcal{I}_2)$
15	1.252	**1.019**	1.071	1.029	1.029	1.132	1.402	$\mathcal{G}_2 \ll \mathcal{I}_1 \ll \mathcal{G}_1 \ll \mathcal{E}_1 \ll (\mathcal{H}_2, \mathcal{H}_1, \mathcal{I}_2)$
20	1.236	**1.018**	1.086	1.022	1.035	1.139	1.486	$\mathcal{G}_2 \ll \mathcal{I}_1 \ll \mathcal{G}_1 \ll \mathcal{E}_1 \ll \mathcal{H}_2 \ll (\mathcal{I}_2, \mathcal{H}_1)$
Avg	1.290	**1.030**	1.068	1.039	1.034	1.128	1.400	

Finally, $\mathcal{E}_1$ performs the next best followed by $\mathcal{G}_1$ that produces solutions whose average relative entropy is 9% worse than those produced by the best scheme.

4.4 Evaluation of k-Way Clustering via Repeated Bisections

Our second set of experiments was focused on evaluating the clustering solutions produced by the various criterion functions when the overall solution was obtained via a sequence of cluster bisections (RB). In this approach, a k-way solution is obtained by first bisecting the entire collection. Then, one of the two clusters is selected and it is further bisected, leading to a total of three clusters. This step of selecting and bisecting a cluster is performed $k - 1$ times leading to the desired k-way clustering solution. Each of these bisections is performed so that the resulting bisection optimizes a particular criterion function. However, the overall k-way clustering solution will not necessarily be at a local optimum with respect to that criterion function.

The key step in this algorithm is the method used to select the cluster to bisect next and a number of different approaches are described in [259, 386, 407]. In all our experiments, we selected the largest cluster, as this approach leads to reasonably good and balanced clustering solutions [407].

The average relative entropies of the resulting solutions are shown in Table 4, and these results are in general consistent with those obtained for direct k-way clustering Table 3. The $\mathcal{I}_1$ and $\mathcal{G}_2$ functions lead to the worst solutions, $\mathcal{H}_2$ leads to the best overall solutions, and $\mathcal{I}_2$, $\mathcal{E}_1$, and $\mathcal{G}_1$ are within 2% of the best. However, in the case of RB, there is a reduction in the relative difference between the best and the worst schemes. For example, $\mathcal{G}_2$ is only 13% worse than the best (compared to 35% for direct k-way). Similar trends can be observed for the other functions. This relative improvement becomes most apparent for the $\mathcal{G}_1$ criterion function that now almost always performs within 2% of the best. The reason for these improvements is discussed in Sect. 5.

Figure 1 compares the quality of the solutions obtained via direct k-way to those obtained via RB. These plots were obtained by dividing the entropies of the solutions obtained by the direct k-way approach with those obtained by the RB approach and averaging them over the 15 datasets. Ratios that are greater than 1 indicate that the RB approach leads to better solutions than

Table 4. Average relative entropies for the clustering solutions obtained via repeated bisections and their statistical significance

k	$\mathcal{I}_1$	$\mathcal{I}_2$	$\mathcal{E}_1$	$\mathcal{H}_1$	$\mathcal{H}_2$	$\mathcal{G}_1$	$\mathcal{G}_2$	Statistical significance test, p-value=0.05
5	1.207	1.050	1.060	1.083	**1.049**	1.053	1.191	$(\mathcal{I}_1, \mathcal{G}_2) \ll (\mathcal{H}_1, \mathcal{E}_1, \mathcal{G}_1, \mathcal{I}_2, \mathcal{H}_2)$
10	1.243	1.112	1.083	1.129	**1.056**	1.106	1.221	$(\mathcal{I}_1, \mathcal{G}_2) \ll (\mathcal{H}_1, \mathcal{I}_2, \mathcal{G}_1, \mathcal{E}_1, \mathcal{H}_2)$
15	1.190	1.085	1.077	1.102	**1.079**	1.085	1.205	$(\mathcal{G}_2, \mathcal{I}_1) \ll (\mathcal{H}_1, \mathcal{G}_1, \mathcal{E}_1, \mathcal{I}_2, \mathcal{H}_2)$
20	1.183	1.070	**1.057**	1.085	1.072	1.075	1.209	$(\mathcal{G}_2, \mathcal{I}_1) \ll (\mathcal{H}_1, \mathcal{G}_1, \mathcal{E}_1, \mathcal{I}_2, \mathcal{H}_2)$
Avg	1.206	1.079	1.069	1.100	**1.064**	1.080	1.207	

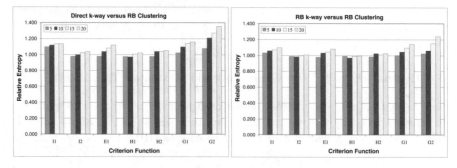

Fig. 1. The relative performance of direct k-way clustering over that of repeated bisections (left). The relative performance of repeated bisections-based clustering followed by k-way refinement over that of repeated bisections alone (right). These results correspond to averages over the different datasets

direct k-way and vice versa. From these plots we see that the direct k-way solutions obtained by $\mathcal{I}_1$, $\mathcal{G}_1$, and $\mathcal{G}_2$ are worse than those obtained by RB clustering. For the remaining functions, the relative performance appears to be sensitive to the number of clusters. For a small number of clusters, the direct approach tends to lead to better solutions; however, as the number of clusters increases the RB approach tends to outperform it. In fact, the sensitivity on k appears to be true for all seven criterion functions, and the main difference has to do with how quickly the relative quality of the direct k-way clustering solution degrades. Among the different functions, $\mathcal{I}_2$, $\mathcal{H}_1$, and $\mathcal{H}_2$ appear to be the least sensitive as their relative performance does not change significantly between the two clustering methods as k increases.

4.5 Evaluation of k-Way Clustering via Repeated Bisections Followed by k-Way Refinement

To further investigate the behavior of the RB-based clustering approach we performed a sequence of experiments in which the final solution obtained by the RB approach for a particular criterion function was further refined using the greedy k-way refinement algorithm described in Sect. 3.2. We refer to this scheme as *RB-k-way*. The average relative entropies for this set of experiments are shown in Table 5.

Comparing the relative performance of the various criterion functions we can see that they are more similar to those of direct k-way (Table 3) than those of the RB-based approach (Table 4). In particular, $\mathcal{I}_2$, $\mathcal{E}_1$, $\mathcal{H}_1$, and $\mathcal{H}_2$ tend to outperform the rest, with $\mathcal{I}_2$ performing the best. Also, we can see that $\mathcal{I}_1$, $\mathcal{G}_1$, and $\mathcal{G}_2$ are considerably worse than the best scheme. Figure 1 compares the relative quality of the RB-k-way solutions to the solutions obtained by the RB-based scheme. Looking at these results we can see that by optimizing the $\mathcal{I}_1$, $\mathcal{E}_1$, $\mathcal{G}_1$, and $\mathcal{G}_2$ criterion functions, the quality of the solutions become worse,

Table 5. Average relative entropies for the clustering solutions obtained via repeated bisections followed by k-way refinement and their statistical significance

k	$\mathcal{I}_1$	$\mathcal{I}_2$	$\mathcal{E}_1$	$\mathcal{H}_1$	$\mathcal{H}_2$	$\mathcal{G}_1$	$\mathcal{G}_2$	Statistical significance test, p-value=0.05
5	1.304	1.081	1.077	1.121	**1.076**	1.097	1.273	$(\mathcal{I}_1,\mathcal{G}_2) \ll (\mathcal{H}_1,\mathcal{G}_1,\mathcal{E}_1,\mathcal{I}_2,\mathcal{H}_2)$
10	1.278	1.065	1.088	1.063	**1.051**	1.127	1.255	$(\mathcal{G}_2,\mathcal{I}_1) \ll \mathcal{G}_1 \ll (\mathcal{E}_1,\mathcal{H}_1,\mathcal{I}_2,\mathcal{H}_2)$
15	1.234	**1.037**	1.089	1.057	1.046	1.140	1.334	$\mathcal{G}_2 \ll \mathcal{I}_1 \ll (\mathcal{G}_1,\mathcal{E}_1) \ll (\mathcal{H}_1,\mathcal{H}_2,\mathcal{I}_2)$
20	1.248	**1.030**	1.098	1.041	1.051	1.164	1.426	$\mathcal{G}_2 \ll \mathcal{I}_1 \ll \mathcal{G}_1 \ll \mathcal{E}_1 \ll (\mathcal{H}_2,\mathcal{H}_1,\mathcal{I}_2)$
Avg	1.266	**1.053**	1.088	1.070	1.056	1.132	1.322	

especially for a large number of clusters. The largest degradation happens for $\mathcal{G}_1$ and $\mathcal{G}_2$. On the other hand, as we optimize either $\mathcal{I}_2$, $\mathcal{H}_1$, or $\mathcal{H}_2$, the overall cluster quality changes only slightly (sometimes it gets better and sometimes it gets worse). These results verify the observations we made in Sect. 4.4 that suggest that the optimization of some of the criterion functions does not necessarily lead to better quality clusters, especially for large values of k.

5 Discussion and Analysis

The experiments presented in Sect. 4 showed two interesting trends. First, the quality of the solutions produced by some seemingly similar criterion functions is often substantially different. For instance, both $\mathcal{I}_1$ and $\mathcal{I}_2$ find clusters by maximizing a particular within cluster similarity function. However, $\mathcal{I}_2$ performs substantially better than $\mathcal{I}_1$. This is also true for $\mathcal{E}_1$ and $\mathcal{G}_1$ that attempt to minimize a function that takes into account both the within cluster similarity and the across cluster dissimilarity. However, in most of the experiments, $\mathcal{E}_1$ tends to perform consistently better than $\mathcal{G}_1$. The second trend is that for many criterion functions, the quality of the solutions produced via RB is better than the corresponding solutions produced either via direct k-way clustering or after performing k-way refinement. Furthermore, this performance gap seems to increase with the number of clusters k. In the remainder of this section we present an analysis that explains the cause of these trends. Our analyses are specific to selected criterion functions, and thus may have limited direct transfer in cases where other criteria are used. However, we believe that such analyses of criteria biases are important generally to better understand empirical findings. This is particularly important in clustering studies, an area in which a plethora of criteria exist, some appearing quite similar in form, but with very different implications for clustering results.

5.1 Analysis of the $\mathcal{I}_1$ and $\mathcal{I}_2$ Criterion Functions

As a starting point for analyzing the $\mathcal{I}_1$ and $\mathcal{I}_2$ criterion functions, it is important to qualitatively understand the solutions that they produce. Figure 2 shows the 10-way clustering solutions obtained for the *sports* dataset using the direct clustering approach for $\mathcal{I}_1$ and $\mathcal{I}_2$. The rows of each subtable represent

cid	Size	Sim	baseball	basketball	football	hockey	boxing	bicycling	golfing
1	1035	0.098	1034	1					
2	594	0.125			1	592			1
3	322	0.191			321	1			
4	653	0.127			1		652		
5	413	0.163	413						
6	1041	0.058				1041			
7	465	0.166			464	1			
8	296	0.172			296				
9	3634	0.020	1393	789	694	157	121	145	335
10	127	0.268	108	1	17			1	

$\mathcal{I}_1$ Criterion Function (Entropy=0.357)

cid	Size	Sim	baseball	basketball	football	hockey	boxing	bicycling	golfing
1	475	0.087	97	35	143	8	112	64	16
2	384	0.129	1	1		381			1
3	1508	0.032	310	58	1055	11	5	59	10
4	844	0.094	1	1	841				1
5	400	0.163		1		399			
6	835	0.097	829			6			
7	1492	0.067	1489		1	2			
8	756	0.099	2	752	1				
9	621	0.108	618	1	2				
10	1265	0.036	65	560	296	9	5	22	308

$\mathcal{I}_2$ Criterion Function (Entropy=0.240)

Fig. 2. The cluster-class distribution of the clustering solutions for the $\mathcal{I}_1$ and $\mathcal{I}_2$ criterion functions for the *sports* dataset

a particular cluster and show the class distribution of the documents assigned to it. The columns labeled "Size" show the number of documents assigned to each cluster and those labeled "Sim" show the average pairwise similarity between the documents of each cluster. From these results we can see that both $\mathcal{I}_1$ and $\mathcal{I}_2$ produce solutions that contain a mixture of large, loose clusters and small, tight clusters. However, $\mathcal{I}_1$ behaves differently from $\mathcal{I}_2$ in two ways: (i) $\mathcal{I}_1$'s solution has a cluster (*cid = 9*), which contains a very large number of documents from different categories and very low average pairwise similarities, whereas $\mathcal{I}_2$'s solution does not. This is also the reason why $\mathcal{I}_1$'s solution has a higher overall entropy value compared to $\mathcal{I}_2$'s (0.357 vs. 0.240). (ii) Excluding this large poor cluster, $\mathcal{I}_1$'s remaining clusters tend to be quite pure and relatively *tight* (i.e., high "Sim" values), whereas $\mathcal{I}_2$'s clusters are somewhat less pure and less tight. The above observations on the characteristics of the solutions produced by $\mathcal{I}_1$ and $\mathcal{I}_2$ and the reasons as to why the former leads to higher entropy solutions hold for the remaining datasets as well.

To analyze this behavior we focus on the properties of an optimal clustering solution with respect to either $\mathcal{I}_1$ or $\mathcal{I}_2$ and show how the tightness of each cluster affects the assignment of documents between the clusters. The following two propositions, whose proofs are in Appendix 6, state the properties that are satisfied by the optimal solutions produced by the $\mathcal{I}_1$ and $\mathcal{I}_2$ criterion functions:

Proposition 1. *Given an optimal k-way solution $\{S_1, S_2, \ldots, S_k\}$ with respect to $\mathcal{I}_1$, then for each pair of clusters S_i and S_j, each document $d \in S_i$ satisfies the following inequality:*

$$\delta_i - \delta_j \geq \frac{\mu_i - \mu_j}{2}, \tag{21}$$

where μ_i is the average pairwise similarity between the document of S_i excluding d, δ_i is the average pairwise similarity between d and the other documents of S_i, μ_j is the average pairwise similarity between the document of S_j, and δ_j is the average pairwise similarity between d and the documents of S_j.

Proposition 2. *Given an optimal k-way solution $\{S_1, S_2, \ldots, S_k\}$ with respect to $\mathcal{I}_2$, then for each pair of clusters S_i and S_j, each document $d \in S_i$ satisfies the following inequality:*

$$\frac{\delta_i}{\delta_j} \geq \sqrt{\frac{\mu_i}{\mu_j}}, \tag{22}$$

where μ_i, μ_j, δ_i, and δ_j is as defined in Proposition 1.

From (21) and (22), we have that if the optimal solution contains clusters with substantially different tightness, then both criterion functions lead to optimal solutions in which documents that are more similar to a tighter cluster are assigned to a looser cluster. That is, without loss of generality, if $\mu_i > \mu_j$, then a document for which δ_i is small will be assigned to S_j, even if $\delta_j < \delta_i$. However, what differentiates the two criterion functions is how small δ_j can be relative to δ_i before such an assignment can take place. In the case of $\mathcal{I}_1$, even if $\delta_j = 0$ (i.e., document d has *nothing* in common with the documents of S_j), d can still be assigned to S_j as long as $\delta_i < (\mu_i - \mu_j)/2$, i.e., d has a relatively low average similarity with the documents of S_i. On the other hand, $\mathcal{I}_2$ will only assign d to S_j if it has a nontrivial average similarity to the documents of S_j ($\delta_j > \delta_i \sqrt{\mu_j/\mu_i}$). In addition, when δ_i and δ_j are relatively small, that is

$$\delta_j < \mu_j \frac{\alpha - 1}{2(\sqrt{\alpha} - 1)} \quad \text{and} \quad \delta_i < \mu_i \frac{\sqrt{\alpha}(\alpha - 1)}{2(\sqrt{\alpha} - 1)}, \quad \text{where} \quad \alpha = \frac{\mu_i}{\mu_j},$$

for the same value of δ_j, $\mathcal{I}_1$ assigns documents to S_j that have higher δ_i values than $\mathcal{I}_2$ does. Of course whether or not such document assignments will happen, depends on the characteristics of the particular dataset, but as long as the dataset has such characteristics, regardless of how $\mathcal{I}_1$ or $\mathcal{I}_2$ are optimized, they will tend to converge to this type of solution.

These observations explain the results shown in Fig. 2, in which $\mathcal{I}_1$'s clustering solution contains nine fairly pure and tight clusters, and a single large and poor-quality cluster. That single cluster acts almost like a *garbage collector*, which attracts all the peripheral documents of the other clusters.

To graphically illustrate this, Fig. 3 shows the range of δ_i and δ_j values for which the movement of a particular document d from the ith to the jth cluster leads to an improvement in either the $\mathcal{I}_1$ or $\mathcal{I}_2$ criterion function. The plots in Fig. 3a were obtained using $\mu_i = 0.10$, $\mu_j = 0.05$, whereas the plot in Fig. 3b were obtained using $\mu_i = 0.20$ and $\mu_j = 0.05$. For both sets of plots $n_i = n_j = 400$ was used. The x-axis of the plots in Fig. 3 corresponds to δ_j, whereas the y-axis corresponds to δ_i. For both cases, we let these average similarities take values between 0 and 1. The various regions in the plots of Fig. 3 are labeled based on whether or not any of the criterion functions will move d to the other cluster, based on the particular set of δ_i and δ_j values.

Looking at these plots we can see that there is a region of small δ_i and δ_j values for which $\mathcal{I}_1$ will perform the move where $\mathcal{I}_2$ will not. These conditions

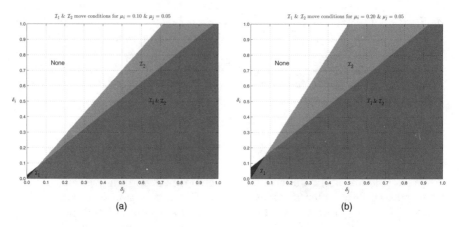

Fig. 3. The range of values of δ_i and δ_j for which a particular document d will move from the ith to the jth cluster. The first plot **(a)** shows the ranges when the average similarities of the documents in the ith and jth cluster are 0.10 and 0.05, respectively. The second plot **(b)** shows the ranges when the respective similarities are 0.20 and 0.05. For both cases each of the clusters was assumed to have 400 documents

are the ones that we already discussed and are the main reason why $\mathcal{I}_1$ tends to create a large poor-quality cluster and $\mathcal{I}_2$ does not. There is also a region for which $\mathcal{I}_2$ will perform the move but $\mathcal{I}_1$ will not. This is the region for which $\delta_i > \delta_j + (\mu_i - \mu_j)/2$ but $\delta_j/\sqrt{\mu_j} > \delta_i/\sqrt{\mu_i}$. That is the average similarity between document d and cluster S_j relative to the square root of the internal similarity of S_j is greater than the corresponding quantity of S_i. Moreover, as the plots illustrate, the size of this region increases as the difference between the tightness of the two clusters increases.

The justification for this type of moves is that d behaves more like the documents in S_j (as measured by $\sqrt{\mu_j}$) than these in S_i. To that extent, $\mathcal{I}_2$ exhibits some dynamic modeling characteristics [260], in the sense that its move is based both on how close it is to a particular cluster as well as on the properties of the cluster itself. However, even though the principle of dynamic modeling has been shown to be useful for clustering, it may sometimes lead to errors as primary evidence of cluster membership (i.e., the actual δ_i and δ_j values) are second guessed. This may be one of the reasons why the $\mathcal{I}_2$ criterion function leads to clusters that in general are more noisy than the corresponding clusters of $\mathcal{I}_1$, as the example in Fig. 2 illustrates.

5.2 Analysis of the $\mathcal{E}_1$ and $\mathcal{G}_1$ Criterion Functions

Both $\mathcal{E}_1$ and $\mathcal{G}_1$ functions measure the quality of the overall clustering solution by taking into account the separation between clusters and the tightness of each cluster. However, as the experiments presented in Sect. 4 show $\mathcal{E}_1$

cid	Size	Sim	baseball	basketball	football	hockey	boxing	bicycling	golfing
1	1330	0.076	1327	2	1				
2	975	0.080	3	5	966				1
3	742	0.072	15	703	24				
4	922	0.079	84	8	32	797			1
5	768	0.078	760	1	6		1		
6	897	0.054	6	2	889				
7	861	0.091	845	0	15				1
8	565	0.079	24	525	13	1			2
9	878	0.034	93	128	114	4	97	121	321
10	642	0.068	255	36	286	7	24	24	10

cid	Size	Sim	baseball	basketball	football	hockey	boxing	bicycling	golfing
1	519	0.146	516		3				
2	597	0.118	1		595				1
3	1436	0.033	53	580	357	13	100	20	313
4	720	0.105		718	1	1			
5	1664	0.032	1387	73	77	49	7	63	8
6	871	0.101	871						
7	1178	0.049	6	5	1167				
8	728	0.111		1		727			
9	499	0.133	498		1				
10	368	0.122	80	33	145	19	15	62	14

$\mathcal{E}_1$ Criterion Function (Entropy=0.203) $\mathcal{G}_1$ Criterion (Entropy=0.239)

Fig. 4. The cluster-class distribution of the clustering solutions for the $\mathcal{E}_1$ and $\mathcal{G}_1$ criterion functions for the *sports* dataset

consistently leads to better solutions than $\mathcal{G}_1$. Figure 4 shows the 10-way clustering solutions produced by $\mathcal{E}_1$ and $\mathcal{G}_1$ for the *sports* dataset and illustrates this difference in the overall clustering quality. As we can see $\mathcal{E}_1$ finds clusters that are considerably more balanced than those produced by $\mathcal{G}_1$. In fact, the solution obtained by $\mathcal{G}_1$ exhibits similar characteristics (but to a lesser extent) with the corresponding solution obtained by the $\mathcal{I}_1$ criterion function described in the previous section. $\mathcal{G}_1$ tends to produce a mixture of large and small clusters, with the smaller clusters being reasonably tight and the larger clusters being quite loose.

In order to compare the $\mathcal{E}_1$ and $\mathcal{G}_1$ criterion functions it is important to rewrite them in a way that makes their similarities and dissimilarities apparent. To this end, let μ_r be the average similarity between the documents of the rth cluster S_r, and let ξ_r be the average similarity between the documents in S_r to the entire set of documents S. Using these definitions, the $\mathcal{E}_1$ and $\mathcal{G}_1$ functions ((15) and (19)) can be rewritten as

$$\mathcal{E}_1 = \sum_{r=1}^{k} n_r \frac{D_r{}^t D}{\|D_r\|} = \sum_{r=1}^{k} n_r \frac{n_r \xi_r}{n_r \sqrt{\mu_r}} = n \sum_{r=1}^{k} n_r \frac{\xi_r}{\sqrt{\mu_r}}, \tag{23}$$

$$\mathcal{G}_1 = \sum_{r=1}^{k} \frac{D_r{}^t (D - D_r)}{\|D_r\|^2} = \left(\sum_{r=1}^{k} \frac{n_r n_r \xi_r}{n_r^2 \mu_r} \right) - k = \left(n \sum_{r=1}^{k} \frac{1}{n_r} \frac{\xi_r}{\mu_r} \right) - k. \tag{24}$$

Note that since k in (24) is constant, it does not affect the overall solution and we will ignore it.

Comparing (23) and (24) we can see that they differ in the way they measure the quality of a particular cluster, and on how they combine these individual cluster quality measures to derive the overall quality of the clustering solution. In the case of $\mathcal{E}_1$, the quality of the rth cluster is measured as $\xi_r/\sqrt{\mu_r}$, whereas in the case of $\mathcal{G}_1$ it is measured as ξ_r/μ_r. Since the quality of each cluster is inversely related to either μ_r or $\sqrt{\mu_r}$, both measures will prefer solutions in which there are no clusters that are extremely loose. Because large clusters tend to have small μ_r values, both the cluster quality measures will tend to produce solutions that contain reasonably balanced clusters. Furthermore, the sensitivity of $\mathcal{G}_1$'s cluster quality measure on clusters with small μ_r values is higher than the corresponding sensitivity of $\mathcal{E}_1$ ($\mu_r \leq \sqrt{\mu_r}$ because $\mu_r \leq 1$). Consequently, we would have expected $\mathcal{G}_1$ to lead to more balanced solutions than $\mathcal{E}_1$, which as the results in Fig. 4 show does not happen, suggesting that the second difference between $\mathcal{E}_1$ and $\mathcal{G}_1$ is the reason for the unbalanced clusters.

The $\mathcal{E}_1$ criterion function sums the individual cluster qualities weighting them proportionally to the size of each cluster. $\mathcal{G}_1$ performs a similar summation but each cluster quality is weighted proportionally to the *inverse* of the size of the cluster. This weighting scheme is similar to that used in the *ratio-cut* objective for graph partitioning [109, 212]. Recall from our previous discussion that since the quality measure of each cluster is inversely related to μ_r, the quality measure of large clusters will have large values, as these clusters will tend to be loose (i.e., μ_r will be small). Now, in the case of $\mathcal{E}_1$, multiplying the quality measure of a cluster by its size ensures that these large loose clusters contribute a lot to the overall value of $\mathcal{E}_1$'s criterion function. As a result, $\mathcal{E}_1$ will tend to be optimized when there are no large loose clusters. On the other hand, in the case of $\mathcal{G}_1$, dividing the quality measure of a large loose cluster by its size has the net effect of decreasing the contribution of this cluster to the overall value of $\mathcal{G}_1$'s criterion function. As a result, $\mathcal{G}_1$ can be optimized at a point in which there exist some large and loose clusters.

5.3 Analysis of the $\mathcal{G}_2$ Criterion Function

The various experiments presented in Sect. 4 showed that the $\mathcal{G}_2$ criterion function consistently led to clustering solutions that were among the worst over the solutions produced by the other criterion functions. To illustrate how $\mathcal{G}_2$ fails, Fig. 5 shows the 10-way clustering solution that it produced via direct k-way clustering on the *sports* dataset. As we can see, $\mathcal{G}_2$ produces solutions that are highly unbalanced. For example, the sixth cluster contains over 2,500 documents from many different categories, whereas the third cluster contains only 42 documents that are primarily from a single category. Note that the clustering solution produced by $\mathcal{G}_2$ is very similar to that produced by the $\mathcal{I}_1$ criterion function (Fig. 2). In fact, for most of the clusters we can find a good one-to-one mapping between the two schemes.

The nature of $\mathcal{G}_2$'s criterion function makes it extremely hard to analyze it. However, one reason that can potentially explain the unbalanced clusters produced by $\mathcal{G}_2$ is the fact that it uses a normalized-cut inspired approach to combine the separation between the clusters (as measured by the cut) versus the size of the respective clusters. It has been shown in [139] that when the normalized-cut approach is used in the context of traditional graph partitioning, it leads to a solution that is considerably more unbalanced than that

cid	Size	Sim	baseball	basketball	football	hockey	boxing	bicycling	golfing
1	491	0.096	1	5	485				
2	1267	0.056	8	5	1244	10			
3	42	0.293	2	1	3		1	35	
4	630	0.113	0	627	2	1			
5	463	0.126	462		1				
6	2596	0.027	1407	283	486	184	42	107	87
7	998	0.040	49	486	124	8	79	3	249
8	602	0.120			1	601			
9	1202	0.081	1194	2	1	5			
10	289	0.198	289						

$\mathcal{G}_2$ Criterion Function (Entropy=0.315)

Fig. 5. The cluster-class distribution of the clustering solutions for the $\mathcal{G}_2$ criterion function for the *sports* dataset

obtained by the $\mathcal{G}_1$ criterion function. However, as our discussion in Sect. 5.2 showed, even $\mathcal{G}_1$'s balancing mechanism often leads to quite unbalanced clustering solutions.

5.4 Analysis of the $\mathcal{H}_1$ and $\mathcal{H}_2$ Criterion Functions

The last set of criterion function that we focus on are the hybrid criterion functions $\mathcal{H}_1$ and $\mathcal{H}_2$, which were derived by combining the $\mathcal{I}_1$ and $\mathcal{E}_1$ and the $\mathcal{I}_2$ and $\mathcal{E}_1$ criterion functions, respectively. The 10-way clustering solutions produced by these criterion functions on the *sports* dataset are shown in Fig. 6. Looking at the results in this table and comparing them against those produced by the $\mathcal{I}_1$, $\mathcal{I}_2$, and $\mathcal{E}_1$, criterion functions we can see that $\mathcal{H}_1$ and $\mathcal{H}_2$ lead to clustering solutions that combine the characteristics of their respective pairs of individual criterion functions. In particular, the $\mathcal{H}_1$ criterion function leads to a solution that is considerably more balanced than that of $\mathcal{I}_1$ and somewhat more unbalanced than that of $\mathcal{E}_1$. Similarly, $\mathcal{H}_2$s solution is also more balanced than $\mathcal{I}_2$ and somewhat less balanced than $\mathcal{E}_1$.

Overall, from the experiments in Sect. 4 we can see that the quality of the solutions (as measured by entropy) produced by $\mathcal{H}_1$ tends to be between that of $\mathcal{I}_1$ and $\mathcal{E}_1$ – but closer to that of $\mathcal{E}_1$s; and the solution produced by $\mathcal{H}_2$ tends to be between that of $\mathcal{I}_2$ and $\mathcal{E}_1$ – but closer to that of $\mathcal{I}_2$s. If the quality is measured in terms of purity, the performance of $\mathcal{H}_1$ relative to $\mathcal{I}_1$ and $\mathcal{E}_1$ remains the same, whereas $\mathcal{H}_2$ tends to outperform both $\mathcal{I}_2$ and $\mathcal{E}_1$.

To understand how these criterion functions, consider the conditions under which a particular document d will move from its current cluster S_i to another cluster S_j. This document will always be moved (or stay where it is), if each one of the two criterion functions used to define either $\mathcal{H}_1$ or $\mathcal{H}_2$ would improve (or degrade) by performing such a move. The interesting case happens when according to one criterion function d should be moved and according to the

cid	Size	Sim	baseball	basketball	football	hockey	boxing	bicycling	golfing
1	1220	0.049	60	20	1131	5	2		2
2	724	0.106		722	1	1			
3	696	0.111			1	694			1
4	1469	0.070	1468	1					
5	562	0.138	560		2				
6	576	0.118	574	1	1				
7	764	0.108	1	1		762			
8	1000	0.045	63	554	370	5	1		7
9	1261	0.023	397	109	130	36	118	145	326
10	308	0.116	289	1	17		1		

$\mathcal{H}_1$ Criterion (Entropy=0.221)

cid	Size	Sim	baseball	basketball	football	hockey	boxing	bicycling	golfing
1	1462	0.997	1457	2	3				
2	908	0.994	2	2	903				1
3	707	0.960	11	679	17				
4	831	0.957	23	4	8	795			1
5	701	0.989	693	1	6		1		
6	999	0.978	15	7	977				
7	830	0.986	818		11				1
8	526	0.949	17	499	7	1			2
9	997	0.321	128	181	149	5	101	113	320
10	619	0.428	248	35	265	8	20	32	11

$\mathcal{H}_2$ Criterion Function (Entropy=0.196)

Fig. 6. The cluster-class distribution of the clustering solutions for the $\mathcal{H}_1$ and $\mathcal{H}_2$ criterion functions for the *sports* dataset

other one d should remain where it is. In that case, the overall decision will depend on how much a particular criterion function improves relative to the degradation of the other function. In general, if such a move leads to a large improvement and a small degradation, it is performed. In order to make such trade-offs possible it is important for the pair of criterion functions involved to take roughly the same range of values (i.e., be of the same order). If that is not true, then improvements in one criterion function will not be comparable to degradations in the other.

In the case of the $\mathcal{H}_1$ and $\mathcal{H}_2$ criterion functions, our studies showed that as long as k is sufficiently large, both the $\mathcal{I}_1$ and $\mathcal{I}_2$ criterion functions are of the same order than $\mathcal{E}_1$. However, in most cases $\mathcal{I}_2$ is closer to $\mathcal{E}_1$ that $\mathcal{I}_1$. This better match between the $\mathcal{I}_2$ and $\mathcal{E}_1$ criterion functions may explain why $\mathcal{H}_2$ seems to perform better than $\mathcal{H}_1$ relative to their respective pairs of criterion functions, and why $\mathcal{H}_1$'s solutions are much closer to those of $\mathcal{E}_1$ instead of $\mathcal{I}_1$.

5.5 Analysis of Direct k-Way Clustering vs. Repeated Bisections

From our analysis of the $\mathcal{I}_1$, $\mathcal{I}_2$, and $\mathcal{G}_1$ criterion functions we know that based on the difference between the tightness (i.e., the average pairwise similarity between the documents in the cluster) of the two clusters, documents that are naturally part of the tighter cluster will end up being assigned to the looser cluster. In other words, the various criterion functions will tend to produce incorrect clustering results when clusters have different degrees of tightness. Of course, the degree to which a particular criterion function is sensitive to tightness differences will be different for the various criterion functions. When the clustering solution is obtained via RB, the difference in tightness between each pair of clusters in successive bisections will tend to be relatively small. This is because each cluster to be bisected, will tend to be relatively homogeneous (due to the way it was discovered), resulting in a pair of subclusters with small tightness differences. On the other hand, when the clustering is

cid	Size	Sim	baseball	basketball	football	hockey	boxing	bicycling	golfing
1	245	0.121	243	0	2				
2	596	0.067	2	1	593				
3	485	0.097	1	480	3	1			
4	333	0.080	3	6	3		2	1	318
5	643	0.104	642		1				
6	674	0.047	669	2	1	1	1		
7	762	0.099	1			760			1
8	826	0.045	42	525	247	6			6
9	833	0.105	832	1					
10	795	0.102	1	1	1	791			1
11	579	0.061	6		573				
12	647	0.034	174	34	156	10	119	144	10
13	191	0.110	189			2			
14	611	0.125	608		3				
15	360	0.168		359	1				

$\mathcal{I}_2$—RB (Entropy=0.125)

cid	Size	Sim	baseball	basketball	football	hockey	boxing	bicycling	golfing
1	292	0.120	280		11		1		
2	471	0.080	1	2	468				
3	468	0.100	1	464	2	1			
4	363	0.072	3	7	5	1	6	20	321
5	545	0.123	542	1	2				
6	1030	0.033	832	36	73	18	4	65	2
7	661	0.110	1	0	660				
8	914	0.046	52	514	334	8	1		5
9	822	0.105	822						
10	771	0.105	1	1		769			
11	641	0.052	2		639				
12	447	0.091	89	30	139	11	110	60	8
13	250	0.105	244		5	1			
14	545	0.138	540		5				
15	360	0.168	2	355	3				

$\mathcal{I}_2$—RB+Renement (Entropy=0.168)

Fig. 7. The cluster-class distribution of the clustering solutions for the $\mathcal{I}_2$ criterion function for the *sports* dataset, for the repeated-bisections solution, and the repeated-bisections followed by k-way refinement

computed directly or when the final k-way clustering obtained via a sequence of RB is refined, there can exist clusters that have significant differences in tightness. Whenever such pairs of clusters occur, most of the criterion functions will end up moving some of the document of the tighter cluster (which are weakly connected to the rest of the documents in that cluster) to the looser cluster. Consequently, the final clustering solution can potentially be worse than that obtained via RB.

To illustrate this behavior we used the $\mathcal{I}_2$ criterion function and computed a 15-way clustering solution using RB and then refined it by performing a 15-way refinement for the *sports* dataset. These results are shown in Fig. 7. The RB solution some clusters that are quite loose and some that are quite tight. Comparing this solution against the one obtained after performing refinement we can see that the size of clusters 6 and 8 (which are among the looser clusters) increased substantially, whereas the size of some of the tighter clusters decreased (e.g., clusters 5, 10, and 14).

6 Conclusions

In this chapter we studied seven different global criterion functions for clustering large documents datasets. Four of these functions ($\mathcal{I}_1$, $\mathcal{I}_2$, $\mathcal{G}_1$, and $\mathcal{G}_2$) have been previously proposed for document clustering, whereas the remaining three ($\mathcal{E}_1$, $\mathcal{H}_1$, and $\mathcal{H}_2$) were introduced by us. Our study consisted of a detailed experimental evaluation using 15 different datasets and three different approaches to find the desired clusters, followed by a theoretical analysis of the characteristics of the various criterion functions. Our experiments showed that $\mathcal{I}_1$ performs poorly whereas $\mathcal{I}_2$ leads to reasonably good results that out-perform the solutions produced by some recently proposed criterion functions ($\mathcal{G}_1$ and $\mathcal{G}_2$). Our three new criterion functions performed reasonably well, with the $\mathcal{H}_2$ criterion function achieving the best overall results.

Our analysis showed that the performance difference observed by the various criterion functions can be attributed to the extent to which the criterion functions are sensitive to clusters of different degrees of tightness, and the extent to which they can lead to reasonably balanced solutions. Moreover, our analysis was able to identify a key property of the $\mathcal{I}_1$ criterion function that can be useful in clustering noisy datasets, in which many documents are segregated to a separate "garbage" cluster.

The various clustering algorithms and criterion functions described in this chapter are available in the CLUTO clustering toolkit, which is available on-line at http://www.cs.umn.edu/~cluto.

Proofs of $\mathcal{I}_1$'s and $\mathcal{I}_2$'s Optimal Solution Properties

Proof (Proposition 1).

For contradiction, let $A_{opt} = \{S_1, S_2, \ldots, S_k\}$ be an optimal solution and assume that there exists a document d and clusters S_i and S_j such that $d \in S_i$ and $\delta_i - \delta_j < (\mu_i - \mu_j)/2$. Consider the clustering solution $A' = \{S_1, S_2, \ldots, \{S_i - d\}, \ldots, \{S_j + d\}, \ldots, S_k\}$. Let D_i, C_i, and D_j, C_j be the composite and centroid vectors of cluster $S_i - d$ and S_j, respectively. Then,

$$
\begin{aligned}
\mathcal{I}_1(A_{opt}) - \mathcal{I}_1(A') &= \frac{\|D_i + d\|^2}{n_i + 1} + \frac{\|D_j\|^2}{n_j} - \left(\frac{\|D_i\|^2}{n_i} + \frac{\|D_j + d\|^2}{n_j + 1} \right) \\
&= \left(\frac{\|D_i + d\|^2}{n_i + 1} - \frac{\|D_i\|^2}{n_i} \right) - \left(\frac{\|D_j + d\|^2}{n_j + 1} - \frac{\|D_j\|^2}{n_j} \right) \\
&= \left(\frac{2n_i d^t D_i + n_i - D_i{}^t D_i}{n_i(n_i + 1)} \right) - \left(\frac{2n_j d^t D_j + n_j - D_j{}^t D_j}{n_j(n_j + 1)} \right) \\
&= \left(\frac{2n_i \delta_i}{n_i + 1} + \frac{1}{n_i + 1} - \frac{n_i \mu_i}{n_i + 1} \right) - \\
&\quad \left(\frac{2n_j \delta_j}{n_j + 1} + \frac{1}{n_j + 1} - \frac{n_j \mu_j}{n_j + 1} \right) \\
&\approx (2\delta_i - 2\delta_j) - (\mu_i - \mu_j),
\end{aligned}
$$

when n_i and n_j are sufficiently large. Since $\delta_i - \delta_j < (\mu_i - \mu_j)/2$, we have $\mathcal{I}_1(A_{opt}) - \mathcal{I}_1(A') < 0$, a contradiction.

Proof (Proposition 2).

For contradiction, let $A_{opt} = \{S_1, S_2, \ldots, S_k\}$ be an optimal solution and assume that there exists a document d and clusters S_i and S_j such that $d \in S_i$ and $\delta_i/\delta_j < \sqrt{\mu_i/\mu_j}$. Consider the clustering solution $A' = \{S_1, S_2, \ldots, \{S_i - $

$d\}, \ldots, \{S_j+d\}, \ldots, S_k\}$. Let D_i, C_i and D_j, C_j be the composite and centroid vectors of cluster $S_i - d$ and S_j, respectively. Then,

$$\mathcal{I}_2(A_{opt}) - \mathcal{I}_2(A') = \|D_i + d\| + \|D_j\| - (\|D_i\| + \|D_j + d\|)$$
$$= (\sqrt{D_i{}^t D_i + 1 + 2d^t D_i} - \sqrt{D_i{}^t D_i}) -$$
$$(\sqrt{D_j{}^t D_j + 1 + 2d^t D_j} - \sqrt{D_j{}^t D_j}). \tag{25}$$

Now, if n_i and n_j are sufficiently large we have that $D_i{}^t D_i + 2d^t D_i \gg 1$, and thus

$$D_i{}^t D_i + 1 + 2d^t D_i \approx D_i{}^t D_i + 2d^t D_i. \tag{26}$$

Furthermore, we have that

$$\left(\sqrt{D_i{}^t D_i} + \frac{d^t D_i}{\sqrt{D_i{}^t D_i}}\right)^2 = D_i{}^t D_i + \frac{(d^t D_i)^2}{D_i{}^t D_i} + 2d^t D_i \approx D_i{}^t D_i + 2d^t D_i, \tag{27}$$

as long as $\delta_i^2 / \mu_i = o(1)$. This condition is fairly mild as it essentially requires that μ_i is sufficiently large relative to δ_i^2, which is always true for sets of documents that form clusters. Now, using (26) and (27) for both clusters, (25) can be rewritten as

$$\mathcal{I}_2(A_{opt}) - \mathcal{I}_2(A') = \frac{d^t D_i}{\sqrt{D_i{}^t D_i}} - \frac{d^t D_j}{\sqrt{D_j{}^t D_j}} = \frac{\delta_i}{\sqrt{\mu_i}} - \frac{\delta_j}{\sqrt{\mu_j}}.$$

Since $\delta_i / \delta_j < \sqrt{\mu_i / \mu_j}$, we have $\mathcal{I}_2(A_{opt}) - \mathcal{I}_2(A') < 0$, a contradiction.

References

1. Available at http://www.cs.umn.edu/~karypis/cluto/files/datasets.tar.gz
2. Available at ftp://ftp.cs.cornell.edu/pub/smart
3. Doc2mat. Available at www-users.cs.umn.edu/karypis/cluto/files/doc2mat-1.0.tar.gz
4. General Text Parser. Available at http://www.cs.utk. edu/lsi/soft.html
5. Latent Semantic Indexing Web Site. Maintained by M.W. Berry and S. Dumais at http://www.cs.utk.edu/lsi/
6. The Lemur Toolkit. Available at http://www-2.cs.cmu.edu/lemur/
7. MATLAB: The Language of Technical Computing. In http://www.mathworks.com/products/matlab/
8. Mc Toolkit. Available at http://www.cs.utexas. edu/users/dml/software/mc/
9. Telcordia Latent Semantic Indexing (LSI) Demo Machine. Available at http://lsi.research.telcordia.com/
10. C.C. Aggarwal, A. Hinneburg, and D.A. Keim. On the surprising behavior of distance metrics in high dimensional space. In *Proceedings of the 8th International Conference on Database Theory (ICDT)*, pages 420–434, London, UK, 2001
11. C.C. Aggarwal, J.L. Wolf, and P.S. Yu. A new method for similarity indexing of market basket data. In *Proceedings ACM SIGMOD International Conference on Management of Data*, pages 407–418, Philadelphia, PA, USA, 1999
12. C.C. Aggarwal, J.L. Wolf, P.S. Yu, C. Procopiuc, and J.S. Park. Fast algorithms for projected clustering. In *Proceedings ACM SIGMOD International Conference on Management of Data*, pages 61–72, Philadelphia, PA, USA, 1999
13. C.C. Aggarwal and P.S. Yu. Finding generalized projected clusters in high dimensional spaces. *Sigmod Record*, 29(2):70–81, 2000
14. R. Agrawal, C. Faloutsos, and A. Swami. Efficient similarity search in sequence databases. In *Proceedings of the 4th International Conference of Foundations of Data Organization and Algorithms (FODO)*, pages 69–84, Evanston, IL, USA, 1993
15. R. Agrawal, J. Gehrke, D. Gunopulos, and P. Raghavan. Automatic subspace clustering of high dimensional data for data mining applications. In *Proceedings of ACM SIGMOD*, pages 94–105, Seattle, Washington, USA, 1998
16. K.S. Al-Sultan. A tabu search approach to the clustering problem. *Pattern Recognition*, 28(9):1443–1451, 1995

17. M.S. Aldenderfer and R.K. Blashfield. *Cluster Analysis*. Number 07-044 in Sage University Paper Series on Quantitative Applications in the Social Sciences. Sage, Beverly Hills, 1984

18. J. Allan. Automatic Hypertext Construction. PhD thesis, Department of Computer Science, Cornell University, January 1995

19. N. Alon, B. Awerbuch, and Y. Azar. The online set cover problem. In *STOC'03: Proceedings of the 35th Annual ACM Symposium on Theory of Computing*, pages 100–105, ACM Press, New York, 2003

20. K. Alsabti, S. Ranka, and V. Singh. An efficient *k*-means clustering algorithm. In *Proceedings of IPPS/SPDP Workshop on High Performance Data Mining*, 1998

21. M.R. Anderberg. *Cluster Analysis and Applications*, Academic, New York, NY, USA, 1973

22. R. Ando. Latent semantic space: iterative scaling improves precision of inter-document similarity measurement. In *Proceedings of the 23rd ACM Conference of SIGIR*, pages 216–223, 2000

23. M. Ankerst, M.M. Breunig, H.-P. Kriegel, and J. Sander. Optics: ordering points to identify the clustering structure. In *Proceedings ACM SIGMOD International Conference on Management of Data*, pages 49–60, Philadelphia, PA, USA, 1999

24. P. Arabie and L.J. Hubert. An overview of combinatorial data analysis. In P. Arabie, L.J. Hubert, and G.D. Soete, editors, *Clustering and Classification*, pages 5–63, World Scientific Publishing Co., Singapore, 1996

25. A.N. Arslan and O.Egecioglu. Efficient algorithms for normalized edit distance. *Journal of Discrete Algorithms*, 1(1), 2000

26. J. Aslam, K. Pelekhov, and D. Rus. Generating, visualizing and evaluating high-quality clusters for information organization. In E.V. Munson, C. Nicholas, and D. Wood, editors, *Principles of Digital Document Processing: 4th International Workshop*, volume 1481 of *Lecture Notes in Computer Science*, pages 53–69, Springer, Berlin Heidelberg New York, 1998

27. J. Aslam, K. Pelekhov, and D. Rus. Static and dynamic information organization with star clusters. In G. Gardarin, J. French, N. Pissinou, K. Makki, and L. Bouganim, editors, *Proceedings of the 7th International Conference on Information and Knowledge Management*, pages 208–217, ACM Press, New York, November 1998

28. J. Aslam, K. Pelekhov, and D. Rus. A practical clustering algorithm for static and dynamic information organization. In *Proceedings of the 10th Annual ACM–SIAM Symposium on Discrete Algorithms*, pages 51–60. ACM SIAM, January 1999

29. J. Aslam, K. Pelekhov, and D. Rus. Information organization algorithms. In *Proceedings of the International Conference on Advances in Infrastructure for Electronic Business, Science, and Education on the Internet*, July 2000

30. J. Aslam, K. Pelekhov, and D. Rus. Using star clusters for filtering. In A. Agah, J. Callan, and E. Rundensteiner, editors, *Proceedings of the 9th International Conference on Information Knowledge Management*, pages 306–313, ACM Press, New York, November 2000

31. J.A. Aslam, E. Pelekhov, and D. Rus. The star clustering algorithm for static and dynamic information organization. *Journal of Graph Algorithms and Applications*, 8(1):95–129, 2004

32. J.A. Aslam, E. Pelekhov, and D. Rus. Persistent queries over dynamic text streams. *International Journal of Electronic Business*, 3(3–4):288–299, 2005

33. J. Aslam, F. Reiss, and D. Rus. Scalable information organization. In *Proceedings of the Conference on Content-Based Multimedia Information Access*, pages 1033–1042, Paris, France, April 2000. CID–CASIS

34. A. Auslender and M. Teboulle. The log-quadratic proximal methodology in convex optimization algorithms and variational inequalities. In P. Daniel, F. Gianessi, and A. Maugeri, editors, *Equilibrium Problems and Variational Models, Nonconvex Optimization and Its Applications*, volume 68, Kluwer, Dordrecht, 2003

35. A. Auslender, M. Teboulle, and S. Ben-Tiba. Interior proximal and multiplier methods based on second order homogeneous kernels. *Mathematics of Operations Research*, 24:645–668, 1999

36. G.P. Babu and M.N. Murty. A near-optimal initial seed value selection in k-means algorithm using a genetic algorithm. *Pattern Recognition Letters*, 14(10):763–769, 1993

37. G.P. Babu and M.N. Murty. Clustering with evolution strategies. *Pattern Recognition*, 27(2):321–329, 1994

38. R. Baeza-Yates. Introduction to data structures and algorithms related to information retrieval. In W.B. Frakes and R. Baeza-Yates, editors, *Information Retrieval, Data Structures and Algorithms*, pages 13–27, Prentice-Hall, Englewood Cliffs, NJ, 1992

39. R. Baeza-Yates, A. Moffat, and G. Navarro. Searching large text collections. In J. Abello, P. Pardalos, and M. Resende, editors, *Handbook of Massive Data Sets*, pages 195–244, Kluwer, Dordrecht, 2002

40. R.A. Baeza-Yates and B.A. Ribeiro-Neto. *Modern Information Retrieval*. New York/Reading, MA, ACM Press/Addison-Wesley, 1999

41. L.D. Baker and A. McCallum. Distributional clustering of words for text classification. In *ACM SIGIR*, pages 96–103, Melbourne, Australia, 1998

42. G. Ball and D. Hall. ISODATA: a novel method of data analysis and pattern classification. Technical report, Stanford Research Institute, Menlo Park, CA, 1965

43. A. Banerjee, I. Dhillon, S. Sra, and J. Ghosh. Generative model-based clustering of directional data. In *Proceedings of the 9th International Conference on Knowledge Discovery and Data Mining (KDD-03)*, pages 19–28, 2003

44. A. Banerjee and J. Ghosh. On scaling up balanced clustering algorithms. In *Proceedings of the 2nd SIAM ICDM*, pages 333–349, Arlington, VA, USA, 2002

45. A. Banerjee and J. Langford. An objective evaluation criterion for clustering. In *Proceedings of the 10th International Conference on Knowledge Discovery and Data Mining (KDD-04)*, pages 515–520, 2004

46. A. Banerjee, S. Merugu, I.S. Dhillon, and J. Ghosh. Clustering with Bregman divergences. In *Proceedings of the 2004 SIAM International Conference on Data Mining*, pages 234–245, SIAM, 2004

47. A. Banerjee, S. Merugu, I.S. Dhillon, and J. Ghosh. Clustering with Bregman divergences. *Journal of Machine Learning Research*, 6:1705–1749, 2005

48. J. Banfield and A Raftery. Model-based gaussian and non-gaussian clustering. *Biometrics*, 49:803–821, 1993

49. D. Barbara and P Chen. Using the fractal dimension to cluster datasets. In *Proceedings of the 6th ACM SIGKDD*, pages 260–264, Boston, MA, USA, 2000

50. J. Becher, P. Berkhin, and E. Freeman. Automating exploratory data analysis for efficient data mining. In *Proceedings of the 6th ACM SIGKDD*, pages 424–429, Boston, MA, USA, 2000

51. N. Beckmann, Kriegel, R.H.-P. Schneider, and B. Seeger. The R^*-tree: An efficient access method for points and rectangles. In *Proceedings of International Conference on Geographic Information Systems*, Ottawa, Canada, 1990

52. D. Beeferman and A. Berger. Agglomerative clustering of a search engine query log. In *Proceedings of the 6th ACM SIGKDD International Conference on Knowledge Discovery and Data Mining*, pages 407–416, 2000

53. M. Belkin and P. Niyogi. Laplacian eigenmaps and spectral techniques for embedding and clustering. In *Advances in Neural Information Processing System*, pages 585–591, The MIT Press, Cambridge, MA, 2002

54. R. Bellman. *Adaptive Control Processes: A Guided Tour*. Princeton University Press, Princeton, NJ, 1961

55. A. Ben-Dor and Z. Yakhini. Clustering gene expression patterns. In *Proceedings of the 3rd Annual International Conference on Computational Molecular Biology (RECOMB99)*, pages 11–14, Lyon, France, 1999

56. A. Ben-Tal, A. Charnes and M. Teboulle. Entropic means. *Journal of Mathematical Analysis and Applications*, 139:537–551, 1989

57. S. Berchtold, C. Böhm, and H.-P. Kriegel. The pyramid-technique: towards breaking the curse of dimensionality. In *Proceedings of the ACM SIGMOD Conference*, pages 142–153, Seattle, WA, USA, 1998

58. P. Berkhin and J.D. Becher. Learning simple relations: theory and applications. In *Proceedings of the 2nd SIAM International Conference on Data Mining*, pages 420–436, Arlington, April 2002

59. P. Berkhin. Survey of clustering data mining techniques. Unpublished manuscript, available at Accrue.com, 2002

60. M. Berry and M. Browne. *Understanding Search Engines*, SIAM, 1999

61. M.W. Berry, S.T. Dumais, and G.W. O'Brien. Using linear algebra for intelligent information retrieval. *SIAM Review*, 37(4):573–595, 1995

62. M. Berry, T. Do, G. O'Brien, V. Krishna, and Sowmini Varadhan. SVDPACKC (Version 1.0) User's Guide. Computer Science Department Technical Report CS-93-194, University of Tennessee, Knoxville, April 1993

63. M.W. Berry, Z. Drmac, and E.R. Jessup. Matrices, vector spaces, and information retrieval. *SIAM Review*, 41(2):335–362, June 1999

64. M.W. Berry. Large scale singular value decomposition. *International Journal of Supercomputer Application*, 6:13–49, 1992

65. M.W. Berry, editor. *Survey of Text Mining: Clustering, Classification, and Retrieval*. Springer, Berlin Heidelberg New York, 2004

66. M.W. Berry, B. Hendrickson, and P. Raghavan. Sparse matrix reordering schemes for browsing hypertext. In J. Renegar, M. Shub, and S. Smale, editors, *The Mathematics of Numerical Analysis*, volume 32 of *Lectures in Applied Mathematics (LAM)*, pages 99–123. American Mathematical Society, Providence, RI, 1996

67. D.P. Bertsekas. *Nonlinear Programming*, Athena Scientific, Belmont, MA, 2nd edition, 1999

68. K. Beyer, J. Goldstein, R. Ramakrishnan, and U. Shaft. When is nearest neighbor meaningful? In *Proceedings of the 7th ICDT*, Jerusalem, Israel, 1999

69. D. Bezdek. *Pattern Recognition with Fuzzy Objective Function Algorithms*. Plenum, New York, NY, USA, 1981

70. C.M. Bishop. *Neural Networks for Pattern Recognition*. Oxford University Press, New York, 1995

71. B.S. Blais. *The Role of the Environment in Synaptic Plasticity: Towards an Understanding of Learning and Memory*. Thesis submitted in partial fulfillment of the requirements for the Degree of Doctor of Philosophy in the Department of Physics at Brown University, http://web.bryant.edu/ bblais/pdf/ chap-introduction.pdf, 1998

72. C. Blake, E. Keogh, and C.J. Merz. UCI repository of machine learning databases. http://www.ics.uci.edu/~mlearn/MLRepository.html, 1998

73. H.H. Bock. Probability models in partitional cluster analysis. In A. Ferligoj and A. Kramberger, editors, *Developments in Data Analysis*, pages 3–25, Slovenia, 1996

74. R.F. Boisvert, R. Pozo, K. Remington, R. Barrett, and J. Dongarra. The matrix market: a web repository for test matrix data. In R.F. Boisvert, editor, *The Quality of Numerical Software, Assessment and Enhancement*, pages 125–137, Chapman and Hall, London, 1997

75. D. Boley, M. Gini, R. Gross, E. Han, K. Hastings, G. Karypis, V. Kumar, B. Mobasher, and J. Moore. Partitioning-based clustering for web document categorization. *Decision Support Systems*, 27:329–341, 1999

76. D.L. Boley. Principal direction divisive partitioning. *Data Mining and Knowledge Discovery*, 2(4):325–344, 1998

77. D.L. Boley. Principal direction divisive partitioning software (experimental software, version 2-beta), February 2003. http://www-users.cs.umn.edu/ boley/Distribution/PDDP2.html

78. B. Bollobás. *Random Graphs*, Academic, London, 1995

79. R.J. Bolton and W.J. Krzanowski. A characterization of principal components for projection pursuit. *The American Statistician*, 53(2):108–109, 1999

80. R.J. Bolton and W.J. Krzanowski. Projection pursuit clustering for exploratory data analysis. *Journal of Computational and Graphical Statistics*, 12(1):121–142, 2003

81. L. Bottou and Y. Bengio. Convergence properties of the K-means algorithms. In G. Tesauro and D. Touretzky, editors, *Advances in Neural Information Processing Systems 7*, pages 585–592, The MIT Press, Cambridge, MA, 1995

82. H. Bozdogan. Determining the number of component clusters in the standard multivariate normal mixture model using model-selection criteria. Technical Report UIC/DQM/A83-1, University of Illinois, Chicago, IL, 1983

83. H. Bozdogan. Mixture-model cluster analysis using model selection criteria and a new information measure of complexity. In *Proceedings of the 1st US/Japan Conference on the Frontiers of Statistical Modeling: An Informational Approach*, pages 69–113, Dordrecht, Netherlands, 1994

84. P. Bradley and U. Fayyad. Refining initial points for k-means clustering. In J. Shavlik, editor, *Proceedings of the 15th International Conference on Machine Learning (ICML), San Francisco, CA*, pages 91–99, AAAI Press, USA, 1998

85. P. Bradley, U. Fayyad, and C. Reina. Scaling clustering algorithms to large databases. In *Proceedings of the 4th International Conference on Knowledge Discovery and Data Mining*. AAAI Press, USA, 1998

86. P.S. Bradley, K.P. Bennett, and A. Demiriz. Constrained k-means clustering. Technical Report MSR-TR-2000-65, Microsoft Research, Redmond, WA, USA, 2000

87. P.S. Bradley, M. Fayyad, and O.L. Mangasarian. Mathematical programming for data mining: formulations and challenges. *INFORMS Journal of Computing*, 11:217–238, 1999

88. L.M. Bregman. A relaxation method of finding a common point of convex sets and its application to the solution of problems in convex programming. *USSR Computational Mathematics and Mathematical Physics.*, 7:200–217, 1967

89. M. Breunig, H.-P. Kriegel, P. Kroger, and J. Sander. Data bubbles: quality preserving performance boosting for hierarchical clustering. In *Proceedings of the ACM SIGMOD Conference*, Santa Barbara, CA, USA, 2001

90. M.M. Breunig, H.-P. Kriegel, R.T. Ng, and J. Sander. Lof: identifying density-based local outliers. In *Proceedings of the ACM SIGMOD Conference*, Dallas, TX, USA, 2000

91. D. Brown and C. Huntley. A practical application of simulated annealing to clustering. Technical Report IPC-TR-91-003, University of Virginia, 1991

92. J. Buhmann. Empirical risk approximation: an induction principle for unsupervised learning. Technical Report IAI-TR-98-3, Institut for Informatik III, Universitat Bonn, 1998

93. S. Busygin, G. Jacobsen, and E. Krämer. Double conjugated clustering applied to leukemia microarray data. In *2nd SIAM ICDM, Workshop on Clustering High Dimensional Data*, Arlington, VA, USA, 2002

94. I. Cadez, S. Gaffney, and P. Smyth. A general probabilistic framework for clustering individuals. Technical Report UCI-ICS 00-09, University of California, Irvine, 2000

95. I. Cadez, P. Smyth, and H. Mannila. Probabilistic modeling of transactional data with applications to profiling, visualization, and prediction. In *Proceedings of the 7th ACM SIGKDD*, pages 37–46, San Francisco, CA, USA, 2001

96. R. Calinski and J. Harabasz. A dendrite method for cluster analysis. *Communications in Statistics*, 3:1–27, 1974

97. G.A. Carpenter, S. Grossberg, and D.B. Rosen. Fuzzy art: fast stable learning and categorization of analog patterns by an adaptive resonance system. *Neural Networks*, 4:759–771, 1991

98. M.A. Carreira-Perpian. A review of dimension reduction techniques. Technical Report CS–96–09, Department of Computer Science, University of Sheffield, January 1997

99. M. Castellanos. Hot-miner: discovering hot topics from dirty text. In Berry [65], pages 123–157

100. G. Celeux and G. Govaert. A classification EM algorithm for clustering and two stochastic versions. *Computational Statistics and Data Analysis,* 14:315, 1992

101. Y. Censor and A. Lent. An interval row action method for interval convex programming. *Journal of Optimization Theory and Applications*, 34:321–353, 1981

102. Y. Censor and S.A. Zenios. *Parallel Optimization,* Oxford University Press, Oxford, 1997

103. S.V. Chakaravathy and J. Ghosh. Scale based clustering using a radial basis function network. *IEEE Transactions on Neural Networks*, 2(5):1250–1261, September 1996

104. M. Charikar, C. Chekuri, T. Feder, and R. Motwani. Incremental clustering and dynamic information retrieval. In *Proceedings of the 29th Symposium on Theory of Computing*, 1997

105. P. Cheeseman and J. Stutz. Bayesian classification (autoclass): theory and results. In U.M. Fayyad, G. Piatetsky-Shapiro, P. Smyth, and R. Uthurusamy, editors, *Advances in Knowledge Discovery and Data Mining*, pages 153–180, AAAI/MIT Press, Cambridge, MA, 1996

106. C. Chen, N. Stoffel, M. Post, C. Basu, D. Bassu, and C. Behrens. Telcordia LSI engine: implementation and scalability issues. In *Proceedings of the 11th Workshop on Research Issues in Data Engineering (RIDE 2001): Document Management for Data Intensive Business and Scientific Applications*, Heidelberg, Germany, April 2001

107. G. Chen and M. Teboulle. Convergence analysis of a proximal-like minimization algorithm using Bregman functions. *SIAM Journal of Optimization*, 3:538–543, 1993

108. C. Cheng, A. Fu, and Y. Zhang. Entropy-based subspace clustering for mining numerical data. In *Proceedings of the 5th ACM SIGKDD*, pages 84–93, San Diego, CA, USA, 1999

109. C.K. Cheng and Y.-C.A. Wei. An improved two-way partitioning algorithm with stable performance. *IEEE Transactions on CAD*, 10:1502–1511, December 1991

110. E. Chisholm and T. Kolda. New term weighting formulas for the vector space method in information retrieval. Report ORNL/TM-13756, Computer Science and Mathematics Division, Oak Ridge National Laboratory, 1999

111. T. Chiu, D. Fang, J. Chen, and Y. Wang. A robust and scalable clustering algorithm for mixed type attributes in large database environments. In *Proceedings of the 7th ACM SIGKDD*, pages 263–268, San Francisco, CA, USA, 2001

112. D. Cook, A. Buja, and J. Cabrera. Projection pursuit indices based on orthonormal function expansions. *Journal of Computational and Graphical Statistics*, 2:225–250, 1993

113. R. Cooley, B. Mobasher, and J. Srivastava. Data preparation for mining world wide web browsing. *Journal of Knowledge Information Systems*, 1(1):5–32, 1999

114. J. Corter and M. Gluck. Explaining basic categories: feature predictability and information. *Psychological Bulletin*, 111:291–303, 1992

115. T.M. Cover and J.A. Thomas. *Elements of Information Theory*, Wiley, New York, USA, 1991

116. M. Craven, D. DiPasquo, D. Freitag, A. McCallum, T. Mitchell, K. Nigam, and S. Slattery. Learning to extract symbolic knowledge from the World Wide Web. In *AAAI98*, pages 509–516, 1998

117. D. Cristofor and D.A. Simovici. An information-theoretical approach to clustering categorical databases using genetic algorithms. In *2nd SIAM ICDM, Workshop on Clustering High Dimensional Data*, Arlington, VA, USA, 2002

118. W.B. Croft. Clustering large files of documents using the single-link method. *Journal of the American Society for Information Science*, pages 189–195, November 1977

119. I. Csiszar. Information-type measures of difference of probability distributions and indirect observations. *Studia Scientiarum Matematicarum Hungar*, 2:299–318, 1967

120. D.R. Cutting, D.R. Karger, J.O. Pedersen, and J.W. Tukey. Scatter/gather: A cluster-based approach to browsing large document collection. In *Proceedings of*

the 15th International ACM SIGIR Conference on Research and Development in Information Retrieval, pages 318–329, 1992

121. C. Daniel and F.C. Wood. *Fitting Equations To Data: Computer Analysis of Multifactor Data*, Wiley, New York, NY, USA, 1980

122. I. Davidson and A. Satyanarayana. Speeding up k-means clustering by bootstrap averaging. In D. Boley et al., editors, *Proceedings of the Workshop on Clustering Large Data Sets (held in conjunction with the Third IEEE International Conference on Data Mining)*, pages 15–25, 2003

123. W. Day and H. Edelsbrunner. Efficient algorithms for agglomerative hierarchical clustering methods. *Journal of Classification*, 1(7):7–24, 1984

124. S. Deerwester, S.T. Dumais, G.W. Furnas, T.K. Landauer, and R. Harshman. Indexing by latent semantic analysis. *Journal of the American Society for Information Science*, 41(6):391–407, 1990

125. D. Defays. An efficient algorithm for a complete link method. *The Computer Journal*, 20:364–366, 1977

126. A. Dempster, N. Laird, and D. Rubin. Maximum likelihood from incomplete data via the EM algorithm. *Journal of the Royal Statistical Society*, 39, 1977

127. J. Devore and R. Peck. *Statistics: The Exploration and Analysis of Data*. Duxbury, Belmont, CA, 1997

128. I.S. Dhillon. Co-clustering documents and words using bipartite spectral graph partitioning. In *Proceedings of the 7th ACM SIGKDD International Conference on Knowledge Discovery and Data Mining (KDD-2001)*, 2001. Also appears as UT CS Technical Report #TR 2001-05, March 1999

129. I.S. Dhillon, J. Fan, and Y. Guan. Efficient clustering of very large document collections. In R. Grossman, C. Kamath, P. Kegelmeyer, V. Kumar, and R. Namburu, editors, *Data Mining for Scientific and Engineering Applications*, pages 357–381, Kluwer, Dordrecht, 2001

130. I.S. Dhillon and Y. Guan. Information-theoretic clustering of sparse co-occurrence data. In *Proceedings of the 2002 IEEE International Conference on Data Mining*, pages 517–520, 2003

131. I.S. Dhillon, Y. Guan, and J. Kogan. Iterative clustering of high dimensional text data augmented by local search. In *Proceedings of the 2002 IEEE International Conference on Data Mining*, pages 131–138, IEEE Computer Society Press, Los Alamitos, CA, 2002

132. I.S. Dhillon, Y. Guan, and J. Kogan. Refining clusters in high-dimensional text data. In I.S. Dhillon and J. Kogan, editors, *Proceedings of the Workshop on Clustering High Dimensional Data and its Applications at the Second SIAM International Conference on Data Mining*, pages 71–82, SIAM, 2002

133. I.S. Dhillon, J. Kogan, and C. Nicholas. Feature selection and document clustering. In M.W. Berry, editor, *A Comprehensive Survey of Text Mining*, pages 73–100, Springer, Berlin Heildelberg New York, 2003

134. I.S. Dhillon, S. Mallela, and R. Kumar. A divisive information-theoretic feature clustering algorithm for text classification. *Journal of Machine Learning Research (JMLR): Special Issue on Variable and Feature Selection*, 3:1265–1287, March 2003

135. I.S. Dhillon and D.S. Modha. A parallel data-clustering algorithm for distributed memory multiprocessors. In M.J. Zaki and C.T. Ho, editors, *Large-Scale Parallel Data Mining, Lecture Notes in Artificial Intelligence, Volume 1759*, pages 245–260, Springer, Berlin Heildelberg New York, 2000. Presented at the 1999 Large-Scale Parallel KDD Systems Workshop, SanDiego, CA.

136. I.S. Dhillon and D.S. Modha. Concept decompositions for large sparse text data using clustering. *Machine Learning*, 42(1):143–175, January 2001. Also appears as IBM Research Report RJ 10147, July 1999

137. I.S. Dhillon, S. Mallela, and R. Kumar. Enhanced word clustering for hierarchical text classification. In *Proceedings of the 8th ACM SIGKDD International Conference on Knowledge Discovery and Data Mining (KDD-2002)*, pages 191–200, 2002

138. P. Diaconis and D. Freedman. Asymptotics of graphical projection pursuit. *Annals of Statistics*, 12:793–815, 1984

139. C. Ding, X. He, H. Zha, M. Gu, and H. Simon. A min–max cut algorithm for graph partitioning and data clustering. In *Proceeding of the IEEE International Conference on Data Mining*, 2001

140. B. Dom. An information-theoretic external cluster-validity measure. Technical report, 2001

141. P. Domingos and G. Hulten. A general method for scaling up machine learning algorithms and its application to clustering. In *Proceedings of the 18th International Converence on Machine Learning*, pages 106–113, Morgan Kaufmann, San Fransisco, USA, 2001

142. R.C. Dubes. Cluster analysis and related issues. In C.H. Chen, L.F. Pau, and P.S. Wang, editors, *Handbook of Pattern Recognition and Computer Vision*, pages 3–32, World Scientific, River Edge, NJ, USA, 1993

143. R.O. Duda and P.E. Hart. *Pattern Classification and Scene Analysis*, Wiley, New York, 1973

144. R.O. Duda, P.E. Hart, and D.G. Stork. *Pattern Classification*, Wiley, New York, 2000

145. S. Dudoit and J. Fridlyand. A prediction-based resampling method for estimating the number of clusters in a dataset. *Genome Biology.*, 3(7), 2002

146. W. DuMouchel, C. Volinsky, T. Johnson, C. Cortes, and D. Pregibon. Squashing flat files flatter. In *Proceedings of the 5th ACM SIGKDD*, pages 6–15, San Diego, CA, USA, 1999

147. L. Engleman and J. Hartigan. Percentage points of a test for clusters. *Journal of the American Statistical Association*, 64:1647–1648, 1969

148. S. Epter, M. Krishnamoorthy, and M. Zaki. Clusterability detection and cluster initialization. In I.S. Dhillon and J. Kogan, editors, *Proceedings of the Workshop on Clustering High Dimensional Data and its Applications at the 2nd SIAM International Conference on Data Mining*, pages 47–58, SIAM, 2002

149. L. Ertoz, M. Steinbach, and V. Kumar. A new shared nearest neighbor clustering algorithm and its applications. In *Workshop on Clustering High Dimensional Data and its Applications at 2nd SIAM International Conference on Data Mining*, 2002

150. L. Ertoz, M. Steinbach, and V. Kumar. Finding clusters of different sizes, shapes, and densities in noisy, high dimensional data. Technical report, 2002

151. M. Ester, A. Frommelt, H.-P. Kriegel, and J. Sander. Spatial data mining: database primitives, algorithms and efficient dbms support. *Data Mining and Knowledge Discovery*, 4(2–3):193–216, 2000

152. M. Ester, H.-P. Kriegel, J. Sander, and X. Xu. A density-based algorithm for discovering clusters in large spatial databases with noise. In *Proceedings of the 2nd ACM SIGKDD*, pages 226–231, Portland, OR, USA, 1996

153. M. Ester, H.-P. Kriegel, and X. Xu. A database interface for clustering in large spatial databases. In *Proceedings of the 1st ACM SIGKDD*, pages 94–99, Montreal, Canada, 1995

154. V. Estivill-Castro and I. Lee. AMOEBA: Hierarchical clustering based on spatial proximity using delaunay diagram. In *Proceedings of the 9th International Symposium on Spatial Data Handling*, Beijing, China, 2000

155. B. Everitt. *Cluster Analysis (3rd ed.)*, Edward Arnold, London, UK, 1993

156. C. Faloutsos and K. Lin. Fastmap: a fast algorithm for indexing, data mining and visualization of traditional and multimedia datasets. In *Proceedings of the ACM SIGMOD International Conference on Management of Data, San Jose, CA*, pages 163–174, ACM, New York, 1995

157. C. Faloutsos, M. Ranganathan, and Y. Manolopoulos. Fast subsequence matching in time-series databases. In *Proceedings of the ACM SIGMOD Conference*, pages 419–429, Minneapolis, MN, 1994

158. C. Faloutsos and K.-I. Lin. FastMap: a fast algorithm for indexing, data-mining and visualization of traditional and multimedia datasets. *SIGMOD Record: Proceedings of the ACM SIGMOD International Conference on Management of Data*, 24(2):163–174, 22–25 May 1995

159. D. Fasulo. An analysis of recent work on clustering algorithms. Technical Report UW-CSE01 -03-02, University of Washington, 1999

160. W. Feller. *An Introduction to Probability Theory and Its Applications, Volume 1*, Wiley, New York, 1968

161. M. Fiedler. A property of eigenvectors of nonnegative symmetric matrices and its application to graph theory. *Czecheoslovak Mathematical Journal*, 25(100):619–633, 1975

162. D. Fisher. Cobweb: knowledge acquisition via conceptual clustering. *Machine Learning*, 2:139–172, 1987

163. D. Fisher. Improving inference through conceptual clustering. In *National Conference on Artificial Intelligence*, pages 461–465, 1987

164. D.H. Fisher. Knowledge acquisition via incremental conceptual clustering. *Machine Learning*, 2:139–172, 1987

165. D.H. Fisher. Iterative optimization and simplification of hierarchical clustering. *Journal of Artificial Intelligence Research*, 4:147–179, 1996

166. E. Forgy. Cluster analysis of multivariate data: efficiency vs. interpretability of classifications. *Biometrics*, 21(3):768, 1965

167. S. Forrest, S.A. Hofmeyr, and A. Somayaji. Computer immunology. *Communications of the ACM*, 40:88–96, 1997

168. A. Foss, W. Wang, and O. Zaane. A non-parametric approach to web log analysis. In *1st SIAM ICDM, Workshop on Web Mining*, pages 41–50, Chicago, IL, USA, 2001

169. E.W. Fowlkes and C.L. Mallows. A method for comparing two hierarchical clusterings. *Journal of American Statistical Association*, 78:553–584, 1983

170. W. Frakes. Stemming algorithms. In W. Frakes and R. Baeza-Yates, editors, *Information Retrieval: Data Structures and Algorithms*, pages 131–160, Prentice Hall, New Jersey, 1992

171. C. Fraley and A. Raftery. How many clusters? Which clustering method? Answers via model-based cluster analysis. *The Computer Journal*, 41(8):578–588, 1998

172. C. Fraley and A. Raftery. MCLUST: Software for model-based cluster and discriminant analysis. Technical Report 342, Department of Statistics, University of Washington, 1999

173. C. Fraley and A.E. Raftery. How many clusters? Which clustering method? Answers via model-based cluster analysis. *The Computer Journal.* http://citeseer.nj.nec.com/fraley98how.htm, 1998

174. C. Fraley. Algorithms for model-based Gaussian hierarchical clustering. *SIAM Journal on Scientific Computing*, 20(1):270–281, 1999

175. J.H. Friedman. Exploratory projection pursuit. *Journal of the American Statistical Association*, 82(397):249–266, 1987

176. J.H. Friedman. An overview of predictive learning and function approximation. In V. Cherkassky, J.H. Friedman, and H. Wechsler, editors, *From Statistics to Neural Networks, Proceedings of the NATO/ASI Workshop*, pages 1–61, Springer, Berlin Heildelberg New York, 1994

177. J.H. Friedman. On bias, variance, 0/1-loss, and the curse-of-dimensionality. *Data Mining and Knowledge Discovery*, 1:55–77, 1997

178. J.H. Friedman and J. W. Tukey. A projection pursuit algorithm for exploratory data analysis. *IEEE Transactions on Computers*, C-23(9):881–890, 1974

179. J. Frilyand and S. Dudoit. Applications of resampling methods to estimate the number of clusters and to improve the accuracy of a clustering method. Statistics Berkeley Technical Report. No 600, 2001

180. K. Fukunaga. *Introduction to Statistical Pattern Recognition.* Academic, New York, 1990

181. Ganti, Gehrke, and Ramakrishnan. CACTUS-clustering categorical data using summaries. In *Proceedings of the 5th ACM SIGKDD*, pages 73–83, San Diego, CA, USA, 1999

182. V. Ganti, R. Ramakrishnan, J. Gehrke, A. Powell, and J. French. Clustering large datasets in arbitrary metric spaces. In *Proceedings of the 15th ICDE*, pages 502–511, Sydney, Australia, 1999

183. M.R. Garey and D.S. Johnson. *Computers and Intractability: A Guide to the Theory of NP-Completeness.* Freeman, New York, 1979

184. J. Gennari, P. Langley, and D. Fisher. Models of incremental concept formation. *Artificial Intelligence*, 40:11–61, 1989

185. A. Gersho and R.M. Gray. *Vector quantization and signal compression.* Kluwer, Dordrecht, 1992

186. A.K. Ghosh, A. Schwartzbard, and M. Schatz. Learning program behavior profiles for intrusion detection. In *Proceedings of the SANS Conference and Workshop on Intrusion Detection and Response*, San Francisco, CA, USA, 1999

187. J. Ghosh. Scalable clustering. In N. Ye, editor, *The Handbook of Data Mining*, pages 247–277, Erlbaum, Mahawah, NJ, 2003

188. D. Gibson, J. Kleinberg, and P. Raghavan. Clustering categorical data: an approach based on dynamic systems. In *Proceedings of the 24th International Conference on Very Large Databases*, pages 311–323, New York, NY, USA, 1998

189. J.R. Gilbert, C. Moler, and R. Schreiber. Sparse matrices in MATLAB: design and implementation. Technical Report CSL 91-4, Xerox Palo Alto Research Center, 1991

190. J.R. Gilbert, C. Moler, and R. Schreiber. Sparse matrices in MATLAB: design and implementation. *SIAM Journal Matrix Analysis and Applications*, 13(1):333–356, 1992

191. J.R. Gilbert and S.-H. Teng. MATLAB mesh partitioning and graph separator toolbox, February 2002. ftp://parcftp.xerox.com/pub/gilbert/meshpartdist.zip

192. J.T. Giles, L. Wo, and M. Berry. GTP (General Text Parser) software for text mining. *Statistical Data Mining and Knowledge Discovery*, pages 455–471, 2003

193. M.A. Gluck and J.E. Corter. Information, uncertainty, and the utility of categories. In *Proceedings of the 7th Annual Conference of the Cognitive Science Society*, pages 283–287, Irvine, CA, 1985, Erlbaum, Mahawah, NJ

194. N. Goharian, A. Chowdhury, D. Grossman, and T. El-Ghazawi. Efficiency enhancements for information retrieval using sparse matrix approach. In *Proceedings of 2000 Parallel and Distributed Processing Techniques and Applications (PDPTA)*, Las Vegas, June 2000

195. N. Goharian, T. El-Ghazawi, and D. Grossman. Enterprise text processing: a sparse matrix approach. In *Proceedings of the IEEE International Conference on Information Techniques on Coding & Computing (ITCC 2001)*, Las Vegas, 2001

196. S. Goil, H. Nagesh, and A. Choudhary. MAFIA: efficient and scalable subspace clustering for very large data sets. Technical Report CPDC-TR-9906-010, Northwestern University, 1999

197. D. Goldberg. *Genetic Algorithms in Search, Optimization, and Machine Learning*. Addison-Wesley, Reading, MA, 1989

198. G.H. Golub and C.F. Van Loan. *Matrix Computations*, 2nd edition, The Johns Hopkins University Press, 1989

199. T.R. Golub, D.K. Slonim, P. Tamayo, C. Huard, M. Gaasenbeek, J.P. Mesirov, H. Coller, M.L. Loh, J.R. Downing, M.A. Caligiuri, C.D. Bloomfield, and E.S. Lander. Molecular classification of cancer: class discovery and class prediction by gene expression monitoring. *Science*, 286:531–537, 1999

200. T.F. Gonzalez. Clustering to minimize the maximum intercluster distance. *Theoretical Computer Science*, 38:293–306, 1985

201. G.J. Gordon et al. Translation of microarray data into clinically relevant cancer diagnostic tests using gene expression ratios in lung cancer and mesothelioma. *Cancer Research*, 62:4963–4967, 2002

202. E. Gose, R. Johnsonbaugh, and S. Jost. *Pattern Recognition and Image Analysis*, Prentice Hall, Englewood Cliffs, NJ, 1996

203. G. Govaert. Simultaneous clustering of rows and columns. *Control and Cybernetics*, 24:437–458, 1995

204. K.C. Gowda and G. Krishna. Agglomerative clustering using the concept of mutual nearest neighborhood. *Pattern Recognition*, 10:105–112, 1978

205. L. Grady and E.L. Schwartz. The graph analysis toolbox: image processing on arbitrary graphs, August 2003. http://eslab.bu.edu/software/graphanalysis/graphanalysis.html

206. M. Gu, H. Zha, C. Ding, X. He, and H. Simon. Spectral relaxation models and structure analysis for k-way graph clustering and bi-clustering. Technical Report CSE-01-007, Pennsylvania State University, 2001

207. S. Guha, R. Rastogi, and K. Shim. CURE: a clustering algorithm for large databases. *ACM SIGMOD International Conference on Management of Data*, pages 73–84, 1998

208. S. Guha, R. Rastogi, and K. Shim. Rock: a robust clustering algorithm for categorical attributes. In *Proceedings of the 15th International Conference on Data Engineering*, 1999

209. S. Guha, R. Rastogi, and K. Shim. CURE: an efficient clustering algorithm for large databases. In *Proceedings of 1998 ACM-SIGMOD International Conference on Management of Data*, pages 73–84, 1998

210. V. Guralnik and G. Karypis. A scalable algorithm for clustering sequential data. In *IEEE ICDM 2001*, Silicon Valley, CA, USA, 2001

211. A. Guttman. *R*-trees: a dynamic index structure for spatial searching. In *Proceedings of the ACM SIGMOD Conference*, pages 47–57, Boston, MA, USA, 1984

212. L. Hagen and A.B. Kahng. New spectral methods for ratio cut partitioning and clustering. *IEEE Transactions on CAD*, 11:1074–1085, 1992

213. L.O. Hall, B. Ozyurt, and J.C. Bezdek. Clustering with a genetically optimized approach. *IEEE Transactions on Evolutionary Computation*, 3(2):103–112, 1999

214. P. Hall. Polynomial projection pursuit. *Annals of Statististics*, 17:589–605, 1989

215. E.H. Han, G. Karypis, V. Kumar, and B. Mobasher. Clustering based on association rule hypergraphs. In *SIGMOD Workshop on Research Issues on Data Mining and Knowledge Discovery (SIGMOD-DMKD'97)*, 1997

216. E.H. Han, D. Boley, M. Gini, R. Gross, K. Hastings, G. Karypis, V. Kumar, B. Mobasher, and J. Moore. WebACE: a web agent for document categorization and exploartion. In *Proceedings of the 2nd International Conference on Autonomous Agents*, May 1998

217. J. Han and M. Kamber. *Data Mining: Concepts and Techniques*, Morgan Kaufmann, San Fransisco, 2001

218. J. Han, M. Kamber, and A.K.H. Tung. Spatial clustering methods in data mining: a survey. In H. Miller and J. Han, editors, *Geographic Data Mining and Knowledge Discovery*, Taylor and Francis, London, 2001

219. P. Hansen and N. Mladenovic J-Means: a new local search heuristic for minimum sum of squares clustering. *Pattern Recognition*, 34:405–413, 2001

220. G. Hardy, J.E. Littlewood, and G. Polya. *Inequalities*, Cambridge University Press, Cambridge, 1934

221. D. Harel and Y. Koren. Clustering spatial data using random walks. In *Proceedings of the 7th ACM SIGKDD*, pages 281–286, San Francisco, CA, USA, 2001

222. J. Hartigan. Statistical theory in clustering. *Journal of Classification*, 2:63–76, 1985

223. J. Hartigan and M. Wong. Algorithm as136: a *k*-means clustering algorithm. *Applied Statistics*, 28:100–108, 1979

224. J.A. Hartigan. *Clustering Algorithms*, Wiley, New York, 1975

225. M.A. Hearst and J.O. Pedersen. Reexamining the cluster hypothesis: Scatter/gather on retrieval results. In *ACM SIGIR*, pages 76–84, 1996

226. J. Heer and E. Chi. Identification of web user traffic composition using multimodal clustering and information scent. In *1st SIAM ICDM, Workshop on Web Mining*, pages 51–58, Chicago, IL, USA, 2001

227. B. Hendrickson and R. Leland. An improved spectral graph partitioning algorithm for mapping parallel computations. *SIAM Journal on Scientific Computing*, 16(2):452–469, 1995

228. W. Hersh, C. Buckley, T.J. Leone, and D. Hickam. OHSUMED: an interactive retrieval evaluation and new large test collection for research. In *SIGIR-94*, pages 192–201, 1994

252   References

229. A. Hinneburg and D. Keim. An efficient approach to clustering large multi-media databases with noise. In *Proceedings of the 4th ACM SIGKDD*, pages 58–65, New York, NY, USA, 1998

230. A. Hinneburg and D. Keim. Optimal grid-clustering: towards breaking the curse of dimensionality in high-dimensional clustering. In *Proceedings of the 25th Conference on VLDB*, pages 506–517, Edinburgh, Scotland, USA, 1999

231. D.S. Hochbaum and D.B. Shmoys. A unified approach to approximation algorithms for bottleneck problems. *Journal of ACM*, 33(3):533–550, 1986

232. M. Hochstenbach. A Jacobi–Davidson type SVD method. *SIAM Journal of Scientific Computing*, 23(2):606–628, 2001

233. K.J. Holzinger and H.H. Harman. *Factor Analysis*. University of Chicago Press, Chicago, 1941

234. Z. Huang. Extensions to the k-means algorithm for clustering large data sets with categorical values. *Data Mining and Knowledge Discovery*, 2(3):283–304, 1998

235. P.J. Huber. Projection pursuit. *Annals of Statistics*, 13:435–475, 1985

236. G. Hulten, L. Spencer, and P. Domingos. Mining time-changing data streams. In *Proceedings of the 7th ACM SIGKDD*, pages 97–106, San Francisco, CA, USA, 2001

237. A. Hyvärinen. New approximations of differential entropy for independent component analysis and projection pursuit. In *Advances in Neural Information Processing Systems*, 10:273–279, 1998

238. E.J. Im and K. Yelick. Optimization of sparse matrix kernels for data mining. University of California, Berkeley, unpublished manuscript, 2001

239. N. Intrator and L. Cooper. Objective function formulation of the bcm theory of visual cortical plasticity. *Neural Networks*, 5:3–17, 1992

240. T.S. Jaakkola and D. Haussler. Exploiting generative models in discriminative classifiers. In M.S. Kearns, S.A. Solla, and D.D. Cohn, editors, *Advances in Neural Information Processing Systems-11*, volume 11, pages 487–493, MIT Press, Cambridge, MA, 1999

241. J.E. Jackson. *A User's Guide To Principal Components*, Wiley, New York, 1991

242. A.K. Jain and R.C. Dubes. *Algorithms for Clustering Data*, Prentice Hall, Englewood Cliffs, NJ, 1988

243. A.K. Jain and P.J. Flynn. Image segmentation using clustering. In *Advances in Image Understanding: A Festschrift for Azriel Rosenfeld*, pages 65–83, IEEE Press, USA, 1966

244. A.K. Jain and J. Mao. Artificial neural networks: a tutorial. *IEEE Computer*, 29(3):31–44, 1996

245. A.K. Jain, M.N. Murty, and Flynn P.J. Data clustering: a review. *ACM Computing Surveys*, 31(3):264–323, 1999

246. N. Jardine and C.J. van Rijsbergen. The use of hierarchic clustering in information retrieval. *Information Storage and Retrieval*, 7:217–240, 1971

247. R.A. Jarvis and E.A. Patrick. Clustering using a similarity measure based on shared nearest neighbors. *IEEE Transactions on Computers*, C-22(11), 1973

248. T. Jebara and T. Jaakkola. Feature selection and dualities in maximum entropy discrimination. In *Proceedings of the 16th UIA Conference*, Stanford, CA, USA, 2000

249. T. Joachims. Text categorization with support vector machines: learning with many relevant features. In *Machine Learning: ECML-98, 10th European Conference on Machine Learning*, pages 137–142, 1998

250. N.L. Johnson, S. Kotz, and N. Balakrishnan. *Continuous Univariate Distributions, volume 2*, Wiley, New York, 1995

251. I.T. Jolliffe. *Principal Component Analysis*. Springer, Berlin Heildelberg New York, 1986

252. M. Jones and R. Sibson. What is projection pursuit? *Journal of the Royal Statistical Society, Series A*, 150:1–36, 1987

253. A. Kalton, P. Langley, K. Wagstaff, and J. Yoo. Generalized clustering, supervised learning, and data assignment. In *Proceedings of the 7th ACM SIGKDD*, pages 299–304, San Francisco, CA, USA, 2001

254. E. Kandogan. Visualizing multi-dimensional clusters, trends, and outliers using star coordinates. In *Proceedings of the 7th ACM SIGKDD*, pages 107–116, San Francisco, CA, USA, 2001

255. H. Kargupta, W. Huang, K. Sivakumar, and E.L. Johnson. Distributed clustering using collective principal component analysis. *Knowledge and Information Systems*, 3(4):422–448, 2001

256. R.M. Karp. Reducibility among combinatorial problems, editors Miller, R.E. and Thatcher, J.W. In *Complexity of Computer Computations*, pages 85–103. Plenum, New York, 1972

257. G. Karypis. CLUTO a clustering toolkit. Technical Report 02-017, Department of Computer Science, University of Minnesota, Minneapolis, MN 55455, August 2002

258. G. Karypis, R. Aggarwal, V. Kumar, and S. Shekhar. Multilevel hypergraph partitioning: application in vlsi domain. In *Proceedings ACM/IEEE Design Automation Conference*, USA, 1997

259. G. Karypis and E.-H. Han. Concept indexing: a fast dimensionality reduction algorithm with applications to document retrieval and categorization. Technical Report TR-00-016, Department of Computer Science, University of Minnesota, Minneapolis, MN, USA, 2000

260. G. Karypis, E.-H. Han, and V. Kumar. CHAMELEON: hierarchical clustering using dynamic modeling. *IEEE Computer*, 32(8):68–75, 1999

261. G. Karypis, E.-H. Han, and V. Kumar. Multilevel refinement for hierarchical clustering. Technical Report 99-020, 1999

262. G. Karypis and V. Kumar. A fast and high quality multilevel scheme for partitioning irregular graphs. *SIAM Journal of Scientific Computation*, 20(1):359–392, 1999

263. G. Karypis, R. Aggarwal, V. Kumar, and S. Shekhar. Multilevel hypergraph partitioning: applications in VLSI domain. In *Proceedings of the Design and Automation Conference*, pages 526–529, 1997

264. R. Kass and A. Raftery. Bayes factors. *Journal of American Statistical Association*, 90:773–795, 1995

265. L. Kaufman and P.J. Rousseeuw. *Finding Groups in Data: An Introduction to Cluster Analysis*. Wiley, New York, 1990

266. M. Kearns, Y. Mansour, and A. Ng. An informationtheoretic analysis of hard and soft assignment methods for clustering. In *Proceedings of the 13th UAI*, pages 282–293, 1997

267. E. Keogh, K. Chakrabarti, S. Mehrotra, and M. Pazzani. Locally adaptive dimensionality reduction for indexing large time series databases. In *Proceedings of the ACM SIGMOD Conference*, Santa Barbara, CA, USA, 2001

268. E. Keogh, K. Chakrabarti, M. Pazzani, and S. Mehrotra. Dimensionality reduction for fast similarity search in large time series databases. *Journal of Knowledge and Information Systems*, 3(3), 2001

269. E. Keogh, S. Chu, and M. Pazzani. Ensemble-index: a new approach to indexing large databases. In *Proceedings of the 7th ACM SIGKDD*, pages 117–125, San Francisco, CA, USA, 2001

270. B.W. Kernighan and S. Lin. An efficient heuristic procedure for partitioning graphs. *The Bell System Technical Journal*, 49(2):291–307, 1970

271. B. King. Step-wise clustering procedures. *Journal of the American Statistical Association*, 69:86–101, 1967

272. J. Kleinberg and A. Tomkins. Applications of linear algebra in information retrieval and hypertext analysis. In *Proceedings of 18th ACM SIGMOD-SIGACT-SIGART Symposium on Principles of Database System*, pages 185–193, ACM Press, New York, 1999

273. E. Knorr and R. Ng. Algorithms for mining distance-based outliers in large datasets. In *Proceedings of the 24h Conference on VLDB*, pages 392–403, New York, NY, USA, 1998

274. E. Knorr, R. Ng, and R.H. Zamar. Robust space transformations for distance-based operations. In *Proceedings of the 7th ACM SIGKDD*, pages 126–135, San Francisco, CA, USA, 2001

275. M. Kobayashi, M. Aono, H. Takeuchi, and H. Samukawa. Matrix computations for information retrieval and major and minor outlier cluster detection. *Journal of Computation and Applied Mathematics*, 149(1):119–129, 2002

276. J. Kogan. Clustering large unstructured document sets. In M.W. Berry, editor, *Computational Information Retrieval*, pages 107–117, SIAM, 2000

277. J. Kogan. Means clustering for text data. In M.W. Berry, editor, *Proceedings of the Workshop on Text Mining at the First SIAM International Conference on Data Mining*, pages 47–54, 2001

278. J. Kogan, C. Nicholas, and V. Volkovich. Text mining with hybrid clustering schemes. In M.W. Berry and W.M. Pottenger, editors, *Proceedings of the Workshop on Text Mining (held in conjunction with the Third SIAM International Conference on Data Mining)*, pages 5–16, 2003

279. J. Kogan, C. Nicholas, and V. Volkovich. Text mining with information–theoretical clustering. *Computing in Science & Engineering*, pages 52–59, November/December 2003

280. J. Kogan, M. Teboulle, and C. Nicholas. The entropic geometric means algorithm: an approach for building small clusters for large text datasets. In D. Boley et al., editor, *Proceedings of the Workshop on Clustering Large Data Sets (held in conjunction with the Third IEEE International Conference on Data Mining)*, pages 63–71, 2003

281. J. Kogan, M. Teboulle, and C. Nicholas. Optimization approach to generating families of k-means like algorithms. In I. Dhillon and J. Kogan, editors, *Proceedings of the Workshop on Clustering High Dimensional Data and its Applications (held in conjunction with the Third SIAM International Conference on Data Mining)*, 2003

282. J. Kogan, M. Teboulle, and C. Nicholas. Data driven similarity measures for k-means like clustering algorithms. *Information Retrieval*, 8:331–349, 2005

283. T. Kohonen. The self-organizing map. *Proceedings of the IEEE*, 9:1464–1479, 1990

284. T. Kohonen. *Self-Organizing Maps*. Springer, Berlin Heidelberg New York, 1995

285. T. Kohonen, S. Kaski, K. Lagus, J. Salojrvi, J. Honkela, V. Paatero, and A. Saarela. Self organization of a massive document collection. *IEEE Transactions on Neural Networks*, 11(3):574–585, 2000

286. E. Kokiopoulou and Y. Saad. Polynomial filtering in latent semantic indexing for information retrieval. In *Proceedings of the 27th ACM SIGIR*, pages 104–111, ACM, New York, 2004

287. E. Kolatch. Clustering algorithms for spatial databases: a survey, 2001

288. T. Kolda and B. Hendrickson. Partitioning sparse rectangular and structurally nonsymmetric matrices for parallel computation. *SIAM Journal on Scientific Computing*, 21(6):2048–2072, 2000

289. T. Kolda and D.O'Leary. A semidiscrete matrix decomposition for latent semantic indexing information retrieval. *ACM Transactions on Information Systems*, 16(4):322–346, 1998

290. T.G. Kolda. Limited-Memory Matrix Methods with Applications. PhD thesis, The Applied Mathematics Program, University of Maryland, College Park, MD, 1997

291. D. Koller and M. Sahami. Toward optimal feature selection. In *Proceedings of the 13th ICML*, pages 284–292, Bari, Italy, 1996

292. V.S. Koroluck, N.I. Portenko, A.V. Skorochod, and A.F. Turbin. *The Handbook on Probability Theory and Mathematical Statistics*. Science, Kiev, 1978

293. H.-P. Kriegel, B. Seeger, R. Schneider, and N. Beckmann. The R^*-tree: an efficient access method for geographic information systems. In *Proceedings International Conference on Geographic Information Systems*, Ottawa, Canada, 1990

294. D. Kroese, R. Rubinstein, and T. Taimre. Application of the cross-entropy method to clustering and vector quantization. *Submitted*, 2004

295. J.B. Kruskal. Toward a practical method which helps uncover the structure of a set of observations by finding the line tranformation which optimizes a new "index of condensation. Statistical Computation, R.C. Milton and J.A. Nelder editors, pages 427–440, 1969

296. W. Krzanowski and Y. Lai. A criterion for determining the number of groups in a dataset using sum of squares clustering. *Biometrics*, 44:23–34, 1985

297. H. Kuhn. The Hungarian method for the assignment problem. *Naval Research Logistics Quarterly*, 2:83–97, 1955

298. S. Kullback and R.A. Leibler. On information and sufficiency. *Journal of Mathematical Analysis and Applications*, 22:79–86, 1951

299. S. Kumar and J. Ghosh. GAMLS: a generalized framework for associative modular learning systems. In *Proceedings of the Applications and Science of Computational Intelligence II*, pages 24–34, Orlando, FL, 1999

300. T. Kurita. An efficient agglomerative clustering algorithm using a heap. *Pattern Recognition*, 24(3):205–209, 1991

301. G. Lance and W. Williams. A general theory of classification sorting strategies. *Computer Journal*, 9:373–386, 1967

302. K. Lang. Newsweeder: learning to filter netnews. In *International Conference on Machine Learning*, pages 331–339, 1995

303. B. Larsen and C. Aone. Fast and effective text mining using linear-time document clustering. In *Proceedings of the 5th ACM SIGKDD*, pages 16–22, San Diego, CA, USA, 1999

304. R.M. Larsen. PROPACK: a software package for the symmetric eigenvalue problem and singular value problems on Lanczos and Lanczos bidiagonalization with partial reorthogonalization. http://soi.stanford.edu/rmunk/PROPACK/

305. C.Y. Lee and E.K. Antonsson. Dynamic partitional clustering using evolution strategies. In *Proceedings of the 3rd Asia-Pacific Conference on Simulated Evolution and Learning*, Nagoya, Japan, 2000

306. E. Lee, D. Cook, S. Klinke, and T. Lumley. Projection pursuit for exploratory supervised classification. Technical Report 04-07, Iowa State University, Humboldt-University of Berlin, University of Washington, February 2004

307. W. Lee and S. Stolfo. Data mining approaches for intrusion detection. In *Proceedings of the 7th USENIX Security Symposium*, San Antonio, TX, USA, 1998

308. R. Lehoucq, D.C. Sorensen, and C. Yang. *Arpack User's Guide: Solution of Large-Scale Eigenvalue Problems With Implicitly Restarted Arnoldi Methods*. SIAM, Philadelphia, 1998

309. T. Leighton and S. Rao. Multicommodity max-flow min-cut theorems and their use in designing approximation algorithms. *Journal of the ACM*, 46(6):787–832, 1999

310. T.A. Letsche and M.W. Berry. Large-scale information retrieval with latent semantic indexing. *Information Sciences*, 100(1–4):105–137, 1997

311. E. Levine and E. Domany. Resampling method for unsupervised estimation of cluster validity. *Neural Computation*, 13:2573–2593, 2001

312. D.D. Lewis. Feature selection and feature extraction for text categorization. In *Proceedings of Speech and Natural Language Workshop*, pages 212–217, Morgan Kaufmann San Mateo, CA, February 1992

313. D. D. Lewis. Reuters-21578 text categorization test collection distribution 1.0. http://www.research.att.com/~lewis, 1999

314. L. Liebovitch and T. Toth. A fast algorithm to determine fractal dimensions by box counting. *Physics Letters*, 141A(8), 1989

315. F. Liese and I. Vajda. *Convex Statistical Distances*. Teubner, Leipzig, 1987

316. D. Lin. An information-theoretic definition of similarity. In *Proceedings of the 15th ICML*, pages 296–304, Madison, WI, USA, 1998

317. D. Littau. Using a Low-Memory Factored Representation to Data Mine Large Data Sets. PhD dissertation, Department of Computer Science, University of Minnesota, 2005

318. D. Littau and D. Boley. Using low-memory representations to cluster very large data sets. In D. Barbará and C. Kamath, editors, *Proceedings of the 3rd SIAM International Conference on Data Mining*, pages 341–345, 2003

319. D. Littau and D. Boley. Streaming data reduction using low-memory factored representations. *Information Sciences, Special Issue on Some Current Issues of Streaming Data Mining*, to appear

320. B. Liu, Y. Xia, and P.S. Yu. Clustering through decision tree construction. SIGMOD-00, 2000

321. H. Liu and R. Setiono. A probabilistic approach to feature selection – a filter solution. In *Proceedings of the 13th ICML*, pages 319–327, Bari, Italy, 1996

322. C. Lund and M. Yannakakis. On the hardness of approximating minimization problems. *Journal of ACM*, 41(5):960–981, 1994

323. J. MacQueen. Some methods for classification and analysis of multivariate observations. In *Proceedings of the 5th Berkeley Symposium on Mathematical Statistics and Probability*, pages 281–296, 1967

324. H. Manilla and D. Rusakov. Decomposition of event sequences into independent components. In *Proceedings of the 1st SIAM ICDM*, Chicago, IL, USA, 2001

325. J. Mao and A.K. Jain. A self-organizing network for hyperellipsoidal clustering (HEC). *IEEE Transactions on Neural Networks*, 7(1):16–29, 1996

326. K.V. Mardia, J.T. Kent, and J.M. Bibby. *Multivariate Analysis*. Academic, San Diego, 1979

327. J.L. Marroquin and F. Girosi. Some extensions of the k-means algorithm for image segmentation and pattern classification. Technical Report A.I. Memo 1390, MIT Press, Cambridge, MA, USA, 1993

328. D. Massart and L. Kaufman. *The Interpretation of Analytical Chemical Data by the Use of Cluster Analysis*. Wiley, New York, NY, 1983

329. A. McCallum, K. Nigam, and L.H. Ungar. Efficient clustering of high-dimensional data sets with application to reference matching. In *Proceedings of the 6th ACM SIGKDD*, pages 169–178, Boston, MA, USA, 2000

330. A.K. McCallum. Bow: a toolkit for statistical language modeling, text retrieval, classification and clustering. http://www.cs.cmu.edu/ mccallum/bow, 1996

331. G. McLachlan and K. Basford. *Mixture Models: Inference and Applications to Clustering*. Dekker, New York, NY, 1988

332. G.J. McLachlan and T. Krishnan. *The EM Algorithm and Extentions*. Wiley, New York, 1996

333. M. Meila and D. Heckerman. An experimental comparison of model-based clustering methods. *Machine Learning*, 42:9–29, 2001

334. M. Meila. Comparing clusterings. Technical Report 417, University of Washington, Seattle, WA, 2002

335. R.S. Michalski and R. Stepp. Learning from observations: conceptual clustering. In *Machine Learning: An Artificial Intelligence Approach*. Morgan Kaufmann, San Mateo, CA, 1983

336. G. Milligan and M. Cooper. An examination of procedures for determining the number of clusters in a data set. *Psychometrika*, 50:159–179, 1985

337. B. Mirkin. *Mathematical Classification and Clustering*. Kluwer, Dordrecht, 1996

338. B. Mirkin. Reinterpreting the category utility function. *Machine Learning*, 42(2):219–228, November 2001

339. N. Mishra and R. Motwani, editors, Special issue: Theoretical advances in data clustering. *Machine Learning*, 56, 2004

340. T.M. Mitchell. *Machine Learning*. McGraw-Hill, New York, 1997

341. D.S. Modha and W. Scott Spangler. Feature weighting in k-means clustering. *Machine Learning*, 52(3):217–237, 2003

342. R.J. Mooney and L. Roy. Content-based book recommending using learning for text categorization. In *Proceedings of the SIGIR-99 Workshop on Recommender Systems: Algorithms and Evaluation*, pages 195–204, 1999

343. A. Moore. Very fast em-based mixture model clustering using multiresolution kd-trees. *Advances in Neural Information Processing Systems*, 11, 1999

344. R. Motwani and P. Raghavan. *Randomized Algorithms*. Cambridge University Press, Cambridge, 1995

345. F. Murtagh. A survey of recent advances in hierarchical clustering algorithms. *Computer Journal*, 26(4):354–359, 1983

346. F. Murtagh. *Multidimensional Clustering Algorithms*. Physica-Verlag, Vienna, Austria, 1985

347. H. Nagesh, S. Goil, and A. Choudhary. Adaptive grids for clustering massive data sets. In *Proceedings of the 1st SIAM ICDM*, Chicago, IL, USA, 2001

348. A.Y. Ng, M.I. Jordan, and Y. Weiss. On spectral clustering: analysis and an algorithm. In *Proceedings Neural Information Processing Systems (NIPS 2001)*, 2001

349. R. Ng and J. Han. Efficient and effective clustering methods for spatial data mining. In *Proceedings of the 20th International Conference on Very Large Data Bases (VLDB)*, pages 144–155, Santiago, Chile, 1994

350. K. Nigam, A. McCallum, S. Thrun, and T. Mitchell. Learning to classify text from labeled and unlabeled documents. In *Proceedings of the 15th National Conference on Artificial Intelligence*, pages 792–799, AAAI Press, USA, 1998

351. S. Nishisato. *Analysis of Categorical Data: Dual Scaling and Its Applications*. University of Toronto, Toronto, Canada, 1980

352. J. Oliver, R. Baxter, and C. Wallace. Unsupervised learning using mml. In *Proceedings of the 13th ICML*, Bari, Italy, 1996

353. C. Olson. Parallel algorithms for hierarchical clustering. *Parallel Computing*, 21:1313–1325, 1995

354. S. Oyanagi, K. Kubota, and A. Nakase. Application of matrix clustering to web log analysis and access prediction. In *Proceedings of the 7th ACM SIGKDD, WEBKDD Workshop*, San Francisco, CA, USA, 2001

355. B. Padmanabhan and A. Tuzhilin. Unexpectedness as a measure of interestingness in knowledge discovery. *Decision Support Systems Journal*, 27(3):303–318, 1999

356. B. Padmanabhan and A. Tuzhilin. Small is beautiful: discovering the minimal set of unexpected patterns. In *Proceedings of the 6th ACM SIGKDD*, pages 54–63, Boston, MA, USA, 2000

357. D. Pelleg and A. Moore. Accelerating exact k-means algorithms with geometric reasoning. In *Proceedings of the 5th ACM SIGKDD*, pages 277–281, San Diego, CA, USA, 1999

358. D. Pelleg and A. Moore. X-means: extending k-means with efficient estimation of the number of clusters. In *Proceedings 17th ICML*, Stanford University, USA, 2000

359. C. Perlich, F. Provost, and J. Simonoff. Tree induction vs. logistic regression: a learning-curve analysis. *Journal of Machine Learning Research (JMLR)*, 4:211–255, 2003

360. G. Piatetsky-Shapiro and C.J. Matheus. The interestingness of deviations. In *Proceedings of the AAAI-94 Workshop on Knowledge Discovery in Databases*, 1994

361. M.F. Porter. The Porter stemming algorithm. www.tartarus.org /martin/ PorterStemmer

362. M.F. Porter. An algorithm for suffix stripping. *Program*, 14:130–137, 1980

363. J. Puzicha, T. Hofmann, and J.M. Buhmann. A theory of proximity based clustering: structure detection by optimization. *PATREC: Pattern Recognition*, 33:617–634, 2000

364. J. Quesada. Creating your own LSA space. In T. Landauer, D. McNamara, S. Dennis, and W. Kintsch, editors, *Latent Semantic Anlysis: A Road to Meaning*. Associates Erlbaum, Mahawah, NJ, In press

365. S. Ramaswamy, R. Rastogi, and K. Shim. Efficient algorithms for mining outliers from large data sets. *Sigmoid Record*, 29(2):427–438, 2000

366. W.M. Rand. Objective criteria for the evaluation of clustering methods. *Journal of the American Statistical Association*, 66:846–850, 1971

367. E. Rasmussen. Clustering algorithms. In W. Frakes and R. Baeza-Yates, editors, *Information Retrieval: Data Structures and Algorithms*, pages 419–442. Prentice Hall, Englewood Cliffs, NJ, 1992

368. R. Rastogi and K. Shim. Scalable algorithms for mining large databases. In Jiawei Han, editor, *KDD-99 Tutorial Notes*. ACM, USA, 1999

369. P. Resnik. Using information content to evaluate semantic similarity in a taxonomy. In *Proceedings of IJCAI-95*, pages 448–453, Montreal, Canada, 1995

370. J. Rissanen. Modeling by shortest data description. *Automatica*, 14:465–471, 1978

371. J. Rissanen. *Stochastic Complexity in Statistical Inquiry*. World Scientific, Singapore, 1989

372. R.T. Rockafellar. *Convex Analysis*. Princeton University Press, Princeton, NJ, 1970.

373. K. Rose, E. Gurewitz, and C.G. Fox. A deterministic annealing approach to clustering. *Pattern Recognition Letters*, 11(9):589–594, 1990

374. V. Roth, V. Lange, M. Braun, and J. Buhmann. A resampling approach to cluster validation. In *COMPSTAT*, http://www.cs.uni-bonn.De/braunm, 2002

375. V. Roth, V. Lange, M. Braun, and J. Buhmann. Stability-based validation of clustering solutions. *Neural Computation*, 16(6):1299–1323, 2004

376. R.Y. Rubinstein. The cross-entropy method for combinatorial and continuous optimization. *Methodology and Computing in Applied Probability*, 2:127–190, 1999

377. G. Salton. *Automatic Text Processing: The Transformation, Analysis, and Retrieval of Information by Computer*. Addison-Wesley, Reading, MA, 1989

378. G. Salton. The SMART document retrieval project. In *Proceedings of the Annual International ACM SIGIR Conference on Research and Development in Information Retrieval*, pages 357–358, 1991

379. G. Salton, J. Allan, and C. Buckley. Automatic structuring and retrieval of large text files. *Communications of the ACM*, 37(2):97–108, 1994

380. G. Salton and C. Buckley. Term-weighting approaches in automatic text retrieval. *Information Processing & Management*, 4(5):513–523, 1988

381. G. Salton and M.J. McGill. *Introduction to Modern Retrieval*. McGraw-Hill, New York, 1983

382. G. Salton, A. Wong, and C.S. Yang. A vector space model for automatic indexing. *Communications of the ACM*, 18(11):613–620, 1975

383. J. Sander, M. Ester, H.-P. Kriegel, and X. Xu. Density-based clustering in spatial databases: The algorithm GDBSCAN and its applications. *Data Mining and Knowledge Discovery*, 2(2):169–194, 1998

384. I. Sarafis, A.M.S. Zalzala, and P.W. Trinder. A genetic rule-based data clustering toolkit. In *Congress on Evolutionary Computation (CEC)*, Honolulu, USA, 2002

385. S. Savaresi and D. Boley. On performance of bisecting k-means and pddp. In *Proceedings of the 1st SIAM ICDM*, Chicago, IL, USA, 2001

386. S.M. Savaresi, D.L. Boley, S. Bittanti, and G. Gazzaniga. Cluster selection in divisive clustering algorithms. In *Proceedings of the 2nd SIAM ICDM*, pages 299–314, Arlington, VA, USA, 2002

387. R. Schalkoff. *Pattern Recognition. Statistical, Structural and Neural Approaches*. Wiley, New York, NY, 1991

388. E. Schikuta. Grid-clustering: a fast hierarchical clustering method for very large data sets. In *Proceedings 13th International Conference on Pattern Recognition Volume 2*, pages 101–105, 1996

389. E. Schikuta and M. Erhart. The bang-clustering system: grid-based data analysis. In *Proceeding of Advances in Intelligent Data Analysis, Reasoning about Data, 2nd International Symposium*, pages 513–524, London, UK, 1997

390. G. Schwarz. Estimating the dimension of a model. *The Annals of Statistics*, 6:461–464, 1978

391. D.W. Scott. *Multivariate Density Estimation*. Wiley, New York, NY, 1992

392. B. Shai. A framework for statistical clustering with a constant time approximation algorithms for k-median clustering. *Proceedings of Conference on Learning Theory, formerly Workshop on Computational Learning Theory, COLT-04, to appear*, 2004

393. R. Shamir and R. Sharan. Algorithmic approaches to clustering gene expression data. In T. Jiang, T. Smith, Y. Xu, and M.Q. Zhang, editors, *Current Topics in Computational Molecular Biology*, pages 269–300, MIT Press, Cambridge, MA, 2002

394. G. Sheikholeslami, S. Chatterjee, and A. Zhang. Wavecluster: a multi-resolution clustering approach for very large spatial databases. In *Proceedings of the 24th Conference on VLDB*, pages 428–439, New York, NY, 1998

395. J. Shi and J. Malik. Normalized cuts and image segmentation. *IEEE Transactions on Pattern Analysis and Machine Intelligence*, 22(8):888–905, August 2000

396. R. Sibson. SLINK: an optimally efficient algorithm for the single link cluster method. *Computer Journal*, 16:30–34, 1973

397. A. Silberschatz and A. Tuzhilin. What makes patterns interesting in knowledge discovery systems. *IEEE Transactions on Knowledge and Data Engineering*, 8(6):970–974, 1996

398. A. Singhal, C. Buckley, M. Mitra, and G. Salton. Pivoted document length normalization. In *ACM SIGIR*, 1996

399. S. Sirmakessis, editor. *Text Mining and its Applications (Results of the NEMIS Launch Conference)*, Springer, Berlin Heidelberg New York, 2004

400. N. Slonim and N. Tishby. Document clustering using word clusters via the Information Bottleneck Method. *Proceedings SIGIR*, pages 208–215, 2000

401. N. Slonim and N. Tishby. The power of word clusters for text classification. In *23rd European Colloquium on Information Retrieval Research (ECIR)*, Darmstadt, 2001

402. P. Smyth. Model selection for probabilistic clustering using cross-validated likelihood. Technical Report ICS Tech Report 98-09, Statistics and Computing, 1998

403. P. Smyth. Probabilistic model-based clustering of multivariate and sequential data. In *Proceedings of the 7th International Workshop on AI and Statistics*, pages 299–304, 1999

404. P.H. Sneath and R.R. Sokal. *Numerical Taxonomy.* Freeman, New York, 1973

405. H. Spath. *Cluster Analysis Algorithms.* Ellis Horwood, Chichester, England, 1980

406. C. Spearman. Footrule for measuring correlations. *British Journal of Psychology,* 2:89–108, July 1906

407. M. Steinbach, G. Karypis, and V. Kumar. A comparison of document clustering techniques. In *Proceedings of the 6th ACM SIGKDD, World Text Mining Conference,* Boston, MA, USA, 2000

408. M. Steinbach, G. Karypis, and V. Kumar. A comparison of document clustering techniques. In *KDD Workshop on Text Mining,* 2000

409. A. Strehl and J. Ghosh. A scalable approach to balanced, high-dimensional clustering of market baskets. In *Proceedings of 17th International Conference on High Performance Computing,* pages 525–536, Bangalore, India, 2000

410. A. Strehl and J. Ghosh. Cluster ensembles – a knowledge reuse framework for combining multiple partitions. *Journal of Machine Learning Research (JMLR),* 3(Dec):583–617, 2002

411. A. Strehl and J. Ghosh. Value-based customer grouping from large retail data-sets. In *Proceedings of the SPIE Conference on Data Mining and Knowledge Discovery, Orlando,* volume 4057, pages 33–42, SPIE, April 2000

412. A. Strehl and J. Ghosh. Relationship-based clustering and visualization for high-dimensional data mining. *INFORMS Journal on Computing,* 15(2):208–230, 2003

413. A. Strehl, J. Ghosh, and R. Mooney. Impact of similarity measures on web-page clustering. In *Proceedings of 17th National Conference on AI: Workshop on AI for Web Search (AAAI 2000),* pages 58–64, AAAI, USA, July 2000

414. C. Sugar and G. James. Finding the number of clusters in a data set: an information theoretic approach. *Journal of the American Statistical Association,* 98:750–763, 2003

415. M. Teboulle. Entropic proximal mappings with application to nonlinear programming. *Mathematics of Operation Research,* 17:670–690, 1992

416. M. Teboulle. On φ-divergence and its applications. In F.Y. Phillips and J. Rousseau, editors, *Systems and Management Science by Extremal Methods – Research Honoring Abraham Charnes at Age 70,* pages 255–273, Kluwer, Norwell, MA, 1992

417. M. Teboulle. Convergence of proximal-like algorithms. *SIAM Journal of Optimization,* 7:1069–1083, 1997

418. M. Teboulle and J. Kogan. Deterministic annealing and a k-means type smoothing optimization algorithm for data clustering. In I. Dhillon, J. Ghosh, and J. Kogan, editors, *Proceedings of the Workshop on Clustering High Dimensional Data and its Applications (held in conjunction with the Fifth SIAM International Conference on Data Mining),* pages 13–22, SIAM, Philadelphia, PA, 2005

419. S. Thomopoulos, D. Bougoulias, and C.-D. Wann. Dignet: an unsupervised-learning clustering algorithm for clustering and data fusion. *IEEE Transactions on Aerospace and Electrical Systems,* 31(1–2):1–38, 1995

420. R. Tibshirani, G. Walther, and T. Hastie. Estimating the number of clusters via the gap statistic. *Journal of Royal Statistical Society B,* 63(2):411–423, 2001

421. N. Tishby, F.C. Pereira, and W. Bialek. The information bottleneck method. In *Proceedings of the 37th Annual Allerton Conference on Communication, Control and Computing*, pages 368–377, 1999

422. W.S. Torgerson. Multidimensional scaling, I: Theory and method. *Psychometrika*, 17:401–419, 1952

423. TREC. Text REtrieval conference. http://trec.nist.gov, 1999

424. J.W. Tukey. *Exploratory Data Analysis*. Addison-Wesley, Reading, MA, 1977

425. A.K.H. Tung, J. Han, L.V.S. Lakshmanan, and R.T. Ng. Constraint-based clustering in large databases. In *Proceedings of the 2001 International Conference on Database Theory (ICDT'01)*, 2001

426. A.K.H. Tung, J. Hou, and J. Han. Spatial clustering in the presence of obstacles. In *Proceedings of the 17th ICDE*, pages 359–367, Heidelberg, Germany, 2001

427. H. Turtle. *Inference Networks for Document Retrieval*. PhD thesis, University of Massachusetts, Amherst, 1990

428. S. van Dongen. A cluster algorithm for graphs. Technical Report INS-R0010, National Research Institute for Mathematics and Computer Science in the Netherlands, Amsterdam, The Netherlands, 2000

429. C.J. van Rijsbergen. *Information Retrieval*, second edition, Butterworths, London, 1979

430. V. Vapnik. *The Nature of Statistical Learning Theory*. Springer, Berlin Heidelberg New York, 1995

431. S. Vempala, R. Kannan, and A. Vetta. On clusterings – good, bad and spectral. In *Proceedings of the 41st Symposium on the Foundation of Computer Science, FOCS*, 2000

432. V. Volkovich, J. Kogan, and C. Nicholas. *k*–means initialization by sampling large datasets. In I. Dhillon and J. Kogan, editors, *Proceedings of the Workshop on Clustering High Dimensional Data and its Applications (held in conjunction with SDM 2004)*, pages 17–22, 2004

433. E.M. Voorhees. Implementing agglomerative hierarchical clustering algorithms for use in document retrieval. *Information Processing and Management*, 22(6):465–476, 1986

434. E.M. Voorhees. The cluster hypothesis revisited. In *Proceedings of the Annual International ACM SIGIR Conference on Research and Development in Information Retrieval*, pages 95–104, 1985

435. C. Wallace and D. Dowe. Intrinsic classification by MML – the Snob program. In *Proceedings of the 7th Australian Joint Conference on Artificial Intelligence*, pages 37–44, Armidale, Australia, 1994

436. C. Wallace and P. Freeman. Estimation and inference by compact coding. *Journal of the Royal Statistical Society, Series B*, 49(3):240–265, 1987

437. W. Wang, J. Yang, and R. Muntz. STING: a statistical information grid approach to spatialdata mining. In *Proceedings of the 23rd Conference on VLDB*, pages 186–195, Athens, Greece, 1997

438. W. Wang, J. Yang, and R.R. Muntz. Pk-tree: a spatial index structure for high dimensional point data. In *Proceedings of the 5th International Conference of Foundations of Data Organization*, USA, 1998

439. W. Wang, J. Yang, and R.R. Muntz. Sting+: an approach to active spatial data mining. In *Proceedings 15th ICDE*, pages 116–125, Sydney, Australia, 1999

440. C.-D. Wann and S.A. Thomopoulos. A comparative study of self-organizing clustering algorithms Dignet and ART2. *Neural Networks*, 10(4):737–743, 1997

441. J.H. Ward. Hierarchical grouping to optimize an objective function. *Journal of the American Statistical Association*, 58:236–244, 1963

442. S. Watanabe. *Knowing and Guessing – A Formal and Quantative Study*. Wiley, New York, 1969

443. P. Willet. Recent trends in hierarchical document clustering: A criticial review. *Information Processing and Management*, 24(5):577–597, 1988

444. I.H. Witten, A. Moffat, and T.C. Bell. *Managing Gigabytes: Compressing and Indexing Documents and Images*. Van Nostrand Reinhold, New York, 1994

445. D.I. Witter and M.W. Berry. Downdating the latent semantic indexing model for conceptual information retrieval. *The Computer Journal*, 41(8):589–601, 1998

446. X. Xu, M. Ester, H.-P. Kriegel, and J. Sander. A distribution-based clustering algorithm for mining large spatial datasets. In *Proceedings of the 14th ICDE*, pages 324–331, Orlando, FL, USA, 1998

447. Y. Yang. An evaluation of statistical approaches to text categorization. *Journal of Information Retrieval*, 1(1/2):67–88, May 1999

448. Y. Yang and J.O. Pedersen. A comparative study on feature selection in text categorization. In *Proceedings of the 14th International Conference on Machine Learning*, pages 412–420, Morgan Kaufmann, San Fransisco, 1997

449. A. Yao. On constructing minimum spanning trees in k-dimensional space and related problems. *SIAM Journal on Computing*, 11(4):721–736, 1982

450. C.T. Zahn. Graph-theoretical methods for detecting and describing gestalt clusters. *IEEE Transactions on Computers*, C-20(1):68–86, January 1971

451. O. Zamir, O. Etzioni, O. Madani, and R.M. Karp. Fast and intuitive clustering of web documents. In D. Heckerman, H. Mannila, D. Pregibon, and R. Uthurusamy, editors, *Proceedings of the 3rd International Conference on Knowledge Discovery and Data Mining (KDD-97)*, page 287, AAAI Press, USA, 1997

452. D. Zeimpekis and E. Gallopoulos. PDDP(l): towards a flexible principal direction divisive partitioning clustering algorithm. In D. Boley em et al, editors, *Proceedings of the Workshop on Clustering Large Data Sets (held in conjunction with the Third IEEE International Conference on Data Mining)*, pages 26–35, 2003

453. D. Zeimpekis and E. Gallopoulos. CLSI: a flexible approximation scheme from clustered term-document matrices. In *Proceedings of the 5th SIAM International Conference on Data Mining*, pages 631–635, Newport Beach, SIAM, CA, 2005

454. H. Zha, C. Ding, M. Gu, X. He, and H. Simon. Spectral relaxation for k-means clustering. In *Neural Information Processing Systems*, volume 14, pages 1057–1064, 2001

455. H. Zha, X. He, C. Ding, H. Simon, and M. Gu. Bipartite graph partitioning and data clustering. In *CIKM*, 2001

456. H. Zha and H.D. Simon. On updating problems in latent semantic indexing. *SIAM Journal on Scientific Computing*, 21(2):782–791, March 2000

457. B. Zhang. Generalized k-harmonic means – dynamic weighting of data in unsupervised learning. In *Proceedings of the 1st SIAM ICDM*, Chicago, IL, USA, 2001

458. G. Zhang, B. Kleyner and M. Hsu. A local search approach to k-clustering. *Technical Report HPL-1999-119*, 1999

459. T. Zhang, R. Ramakrishnan, and M. Livny. BIRCH: an efficient data clustering method for very large databases. In *Proceedings of the ACM SIGMOD Conference on Management of Data*, Montreal, Canada, 1996

460. T. Zhang, R. Ramakrishnan, and M. Livny. BIRCH: a new data clustering algorithm and its applications. *Journal of Data Mining and Knowledge Discovery*, 1(2):141–182, 1997

461. Y. Zhang, A.W. Fu, C.H. Cai, and P.-A. Heng. Clustering categorical data. In *Proceedings of the 16th ICDE*, page 305, San Diego, CA, USA, 2000

462. Y. Zhao and G. Karypis. Criterion functions for document clustering: experiments and analysis. Technical Report CS Department 01-40, University of Minnesota, 2001

463. Y. Zhao and G. Karypis. Empirical and theoretical comparisons of selected criterion functions for document clustering. *Machine Learning*, 55(3):311–331, 2004

464. S. Zhong and J. Ghosh. A comparative study of generative models for document clustering. *Knowledge and Intelligent Systems*, 2005

465. D. Zuckerman. NP-complete problems have a version that's hard to approximate. In *Proceedings of the 8th Annual Structure in Complexity Theory Conference*, pages 305–312, IEEE Computer Society, Los Alamitos, CA, 1993

Index

Printing: Krips bv, Meppel
Binding: Stürtz, Würzburg